FUNDAMENTALS OF COMPUTER NUMERICAL CONTROL

Third Edition

FUNDAMENTALS OF COMPUTER NUMERICAL CONTROL

Third Edition

William W. Luggen

Delmar Publishers Inc.™

I(T)P™

NOTICE TO THE READER

A very special thanks to my wife, Linda, for her patience and time spent helping to produce this text. It is to her and my beautiful daughters that I dedicate this book. Most of all, I thank God for the opportunities I've been given.

Cover photo courtesy of Surfware Inc.

Delmar staff

Senior acquisitions editor: Vernon R. Anthony
Editorial assistant: Alison Foster
Project editor: Elena M. Mauceri
Production coordinator: Karen Smith

For information, address Delmar Publishers Inc.
3 Columbia Circle , Box 15–015
Albany, New York 12203–5015

COPYRIGHT © 1994
BY DELMAR PUBLISHERS INC.

The trademark ITP is used under license.

Printed in the United States of America
Published simultaneously in Canada
by Nelson Canada,
A division of The Thomson Corporation

10 9 8 7 6 5

Library of Congress Cataloging-in-Publication Data
Luggen, William W., 1947–
    Fundamentals of computer numerical control / William W.
Luggen.Congresses. – 3rd ed.
        p.   cm.
    Rev. ed. of: Fundamentals of numerical control. 2nd ed., c1988.
    Includes index.
    ISBN 0-8273-6496-2
    1. Machine-tools-Numerical control.   I. Luggen, William W.,
1947– Fundamentals of numerical control.   II. Title.
TJ1189.L84   1994
621.9′023–dc20   93-30141
                                            CIP

CONTENTS

NEW AND REVISED TITLES

Computer Numerical Control: From Programming to Networking/ Lin
ISBN: 0-8273-4715-4

Learning Computer Numerical Control/ Janke
ISBN: 0-8273-4536-4

Computer Numerical Control: Concepts and Programming, 2E/ Seames
ISBN: 0-8273-3782-5

*To request more information on these publications, contact
your local bookstore, or call or write to:*

Delmar Publishers Inc.
3 Columbia Circle
P. O. Box 15015
Albany, NY 12212-5015

Phone: 1-800-347-7707 · 1-518-464-3500 · Fax: 1-518-464-0301

PREFACE

Numerically controlled machine tools have become standard fixtures in the manufacturing facilities of today. Since these tools first began making their way into industry around 1957, their capabilities have greatly increased. The evolution of the field of numerical control has introduced industry to such concepts as CAD/CAM, flexible manufacturing systems, group technology, and many others.

The first edition of the very popular *FUNDAMENTALS OF NUMERICAL CONTROL* was written to provide students with a detailed fundamental look at numerical control from a conceptual point of view. The second edition of *FUNDAMENTALS OF NUMERICAL CONTROL* provided a look forward with new N/C and CNC concepts and functionality, as well as expanded information.

This, the third edition, now brings about a name change to *FUNDAMENTALS OF COMPUTER NUMERICAL CONTROL* in addition to major revision changes and updates to make it the most well-written, up-to-date CNC text on the market today.

As you use the third edition you will notice changes in the following areas:

- Expanded information about tooling, inserts, wear, and chip control
- Over 90 new photographs and illustrations
- Updated information about turning centers and machining centers
- New CNC, DNC, CIM, and CAD/CAM information
- Enhanced programming chapters
- A more in-depth look at flexible manufacturing cells and systems and the future of CNC
- Expanded coverage of language- and graphics-based programming
- More up-to-date examples and explanations throughout the text

William W. Luggen, the author, is Manager, Manufacturing Systems Development for GE Aircraft Engines. His extensive manufacturing engineering experience and his experience as a teacher of N/C programming enable him to write with the needs of both industry and education in mind.

Special features that made the first and second editions of *FUNDAMENTALS OF NUMERICAL CONTROL* so special have been included again in this new third edition.

- Organized, friendly, and easy-to-read writing style
- Subjective review questions to promote original thinking rather than mere memorization
- An appendix of useful formulas and tables to use while studying the chapters to which they pertain
- Updated glossary for definition of technical and key terms used in the text
- Appendix of preparatory functions, miscellaneous functions, and other address characters for quick and easy reference
- Expanded list of safety rules to follow for the use of N/C machines

The first four chapters provide the student with a concise history of numerical control and a description of its real purpose and basic operation. Decimal point programming examples have been added and some older programming examples retained in order to introduce students to the diversity of CNC formats in use today. Also included are concepts of basic coding systems, DNC and communication requirements, and CNC features. These chapters serve as a base for the rest of the text.

Subsequent chapters emphasize concepts and fundamentals, first in a general format and then in more specific formats of turning and machining centers, with coverage of new features, concepts, and capabilities.

The coverage of computer numerical control includes a discussion of language-based as well as graphics-based programming, extending to advanced applications. Tooling is discussed as well, primarily because it is the most underrated aspect of numerical control programming and because of the continuous advances being made in the cutting tool industry. Specific details of setup and fixtures have purposely been omitted; they are a study in themselves and will invariably differ from one manufacturer to another. Extensive calculations have not been included so as to concentrate more thoroughly on N/C programming principles, concepts, and modern features.

As global competition continues to intensify, manufacturing's survival depends on reducing cost and lead time, improving quality and productivity, and maintaining a competent and skilled work force. *FUNDAMENTALS OF COMPUTER NUMERICAL CONTROL* is a small contribution toward achieving those goals.

ACKNOWLEDGMENTS

The author extends his sincere appreciation to Mr. Tim Borkowski of Northeast Metro Technical College for some of the essential materials used throughout this text and for his critical review. Without his help, much of the new material would not have been possible.

The author and publisher recognize with appreciation the technical assistance and photographs that the following companies provided:

CG Tech

Cincinnati Milacron Inc.

Cleveland Twist Drill

DeVlieg-Bullard Services Group

DoAll Company

Gibbs and Associates

Giddings & Lewis Measurement Systems

ITW Woodworth

Jet Edge Corporation

Kennametal, Inc.

Maho Machine Tool Corporation

Mastercam/CNC Software, Inc.

Mazak Corporation

Monarch Machine Tool Company

Murata-Wiedemann, Inc.

Numeridex CAM Division, Automation Intelligence, Inc.

OKuma Corporation

Precision Industries, Sharpaloy Division

Qu-Co Modular Fixturing

Surfware Incorporated

Thomson™ Industries, Inc.

Valenite, Inc.

For their critical comments and recommendations, the author is indebted to the following reviewers:

Mr. Tim Borkowski
Instructor, Machine Technology
Northeast Metro Technical College
White Bear Lake, Minnesota

Robert Brown
Central Carolina Community College
Stanford, NC

Ronald Way
El Camino College
Torrence, CA

Don Creger
Western Illinois University
Macomb, IL

CHAPTER 1

Numerical Control:
History and Evolution

OBJECTIVES

After studying this chapter, you will be able to:

- Discuss the general history of numerical control (N/C) and computer numerical control (CNC)
- Identify some basic types of N/C and CNC machines
- List principal differences between N/C and CNC
- Discuss the general meanings of the terms *accuracy*, *repeatability*, and *reliability* as applied to N/C and CNC equipment

NUMERICAL CONTROL: ONCE UPON A TIME

Perhaps the first real numerical control (N/C) machine was a 1725 knitting machine that was controlled by sheets of punched cardboard. Then, around 1863, the player piano was introduced. The knitting machine and the player piano are considered the true forerunners of modern numerical control. Actually, numerical control as we know it began in 1947. John Parsons of the Parsons Corporation, based in Traverse City, Michigan, began experimenting with the idea of generating thru-axis curve data and using that data to control machine tool motions. Numerical control originated when Parsons discovered a way to couple computer equipment with a jig borer.

In 1949 there was a demand for increased productivity by the U.S. Air Materiel Command as the parts for its airplanes and missiles became more complex. In addition, the designs were constantly being changed and revised. A contract was granted to the Parsons Corporation to search for a speedy production method. In 1951 the Massachusetts Institute of Technology (MIT) assumed the effort. In 1952 MIT successfully demonstrated a primitive model of the N/C machine of today. The machine was a vertical spindle Cincinnati Hydrotel with a lab-constructed control unit. The machine successfully made parts with simultaneous thru-axis cutting tool movements. Actually, MIT coined the term *numerical control*.

In 1955, at the national machine tool show, commercial models of numerically controlled machines were displayed, ready for customer acceptance. Those shown were different contour-milling machines, which cost several hundred thousands of dollars. Some machines required trained and skilled mathematicians and computers to produce tapes. By 1957 numerically controlled machines were accepted by industry; several had been installed and were in use in some production applications.

Machine tool technology has rapidly evolved since the introduction of numerical control. Capabilities have dramatically increased while machine tool and control unit size and cost have decreased. Machines that used to cost $100,000, for example, can now be bought for $30,000 or less. The advancing technologies of precision mechanics and sophisticated electronics have pushed manufacturing efficiencies further and further toward safe, easy-to-use, unattended operation of CNC machine tools. Digital servo systems, high-speed spindles, and synchronous control systems have all combined to boost performance capabilities well beyond machining parameters considered nearly unattainable just a few years ago. In one day it is now possible to take delivery of a CNC machine tool, set it up and produce parts within tolerance and without problems. This is due to reliable electronics, user-friendly control panels and sophisticated software, and advanced diagnostics that pinpoint problems when they occur.

TYPES OF MACHINES CONTROLLED BY N/C AND CNC

At present there are many types of N/C and CNC machines producing parts in manufacturing plants. They range from the earlier models of N/C machine tools (see figure 1-1) to the advanced CNC profilers (see figure 1-2). The sizes, capabilities, and options vary with each N/C or CNC machine. The one common factor is that they are all N/C or CNC controlled and are programmed.

The most common types of N/C machines currently in use are of the chip-making variety, such as N/C turning and machining centers (see figures 1-3 and 1-4). Many of these are now used in toolroom applications as well. Other types of numerical control machines include drafting machines, hole-punching machines, tube benders, inspection machines, riveting machines, welding machines, flame cutters, and precision centerless and centertype grinders. Some examples are shown in figures 1-5, 1-6, and 1-7.

The application of CNC to grinding offers a quick insight into the expanding capabilities of the technology to the machine tool industry. Modern metal removal by abrasion is typically used to rapidly and precisely grind dimensions to the last ten-thousandths of an inch to attain fine, extra-smooth surface finishes on metal already hardened.

Today, manufacturers are finding it faster and more economical to skip turning, heat-treating, and deburring operations. They're starting with a solid blank or near-net-shape material and grinding it to final size and finish in one operation. Grinding finished parts from solid reduces cost per piece and in-process inventory, and it streamlines door-to-door time. For example, consider grinding wheels. Twenty or so years ago they were a static technology. Since then, new abrasives and new bonds have combined to give users a major leap in wheel quality and productivity improvements.

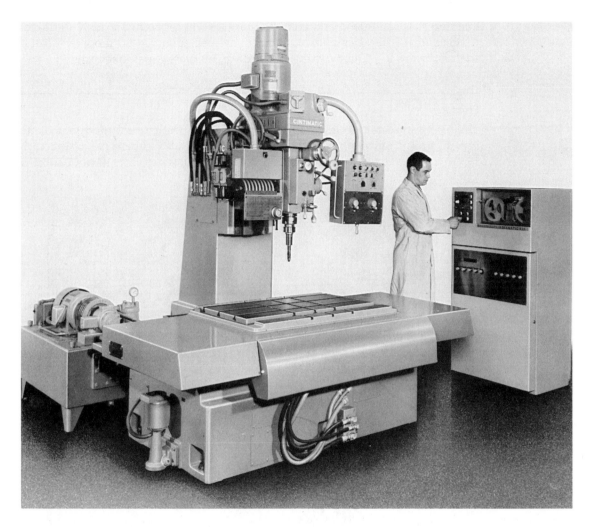

FIGURE 1-1
Early model of an N/C machine. (Courtesy of Cincinnati Milacron Inc.)

The accuracy of CNC grinding is becoming increasingly important as more manufacturers strive to make Six Sigma, or 99.9997% defect-free, products (3.4 defects per million in a product line). Critical to the growth of CNC grinding is the CNC unit's ability to monitor and/or influence some 34 variables, including work speed, coolant flow rate, porosity of the wheel, and harmonic vibration. Continued growth in complex, varied part shapes will push CNC grinding capabilities to ever-expanding parameters far beyond the machine and control unit's present capacity.

For the machine tool industry in general, new changes are affecting the way parts are manufactured. These changes include more near-net shapes, forgings, castings, powdered metal, etc.—all requiring less machining but tighter tolerances. Faster speed and feed rates are also necessitating greatly improved CNC control accuracies and capabilities, as well as increased modular off-the-shelf software to reduce cost for expanding integrated manufacturing.

N/C TO CNC

There is no greater evolutionary aspect of numerical control than that of the control unit. Controls have progressed from the bulky tube types of the early 1950s to the microprocessor-based CNC units of today. The introduction of

FIGURE 1-2
An advanced CNC profiler. (Courtesy of Cincinnati Milacron Inc.)

FIGURE 1-3
A modern CNC turning center. (Courtesy of Okuma Machinery Inc.)

FIGURE 1-4
A typical CNC machining center. (Courtesy of Cincinnati Milacron Inc.)

solid-state circuitry and eventually modular or integrated circuits (ICs) led the way in this electronic revolution and enabled machine tool manufacturers to combine many individual circuits into a single microchip and to offer features and reliability to N/C machines previously unattainable. As a result, control capability and reliability increased as size and cost decreased.

Since the beginning of N/C, approximately 1954, until the early 1970s, all machine control units (MCUs) were *hard-wired*. This meant that all of the logic was fixed—built-in and determined by the physical electronic elements of the control unit. These elements then controlled all functions, such as tape format, absolute or incremental positioning, and character code recognition. In the early 1970s, more capable and less expensive electronics began to emerge. These types of computer elements, or complete microprocessors, became part of the control units. Functions that were solely the result of hardware design became resident in computer logic within the control unit.

FIGURE 1-5
A CNC hole-punching machine. (Courtesy of Wiedemann Division, The Warner & Swasey Co., subsidiary of Bendix Corporation)

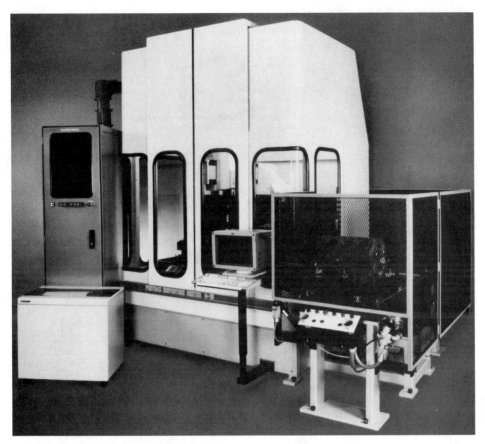

FIGURE 1-6
A CNC coordinate measuring inspection machine. (Courtesy of Giddings and Lewis Measurement Systems)

The totality of software-based controls were first displayed and demonstrated at the 1976 machine tool show. What had once been the result of hardware design was replaced with complete computer logic, which had more capability, was no more costly, and could be programmed for a variety of functions at any time. Essentially, the ability of a machine control unit to recognize different tape formats was not locked in at the time of manufacture. It was simply a matter of how the computer element within the control unit was programmed to read the various tape codes, functions, etc.

The physical components of the software-based CNC units are the same regardless of machine type. It is not the control unit elements, but rather the *executive program,* or load tape, that makes a control unit "think" like a machining center or lathe. Thus, the basic functioning of the CNC unit can be altered by changing the executive, or load program. This executive program is supplied by the control unit manufacturers. In most cases the user does not attempt to alter the executive program in any way.

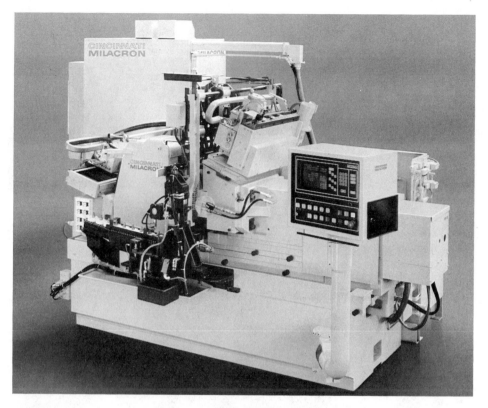

FIGURE 1-7
A CNC centerless grinder. (Courtesy of Cincinnati Milacron Inc.)

The executive program can be revised, updated, or modified at any time. New functionality can be introduced by simply reading them into the executive portion of the control unit computer. The input medium for this process can be punched tape, cassette tape, floppy disk, or a programmable controller unit. Changing functions in a hard-wired control would involve changing the particular controlling elements within the control unit structure.

The advent of computer numerical control (CNC) in the mid-1970s has resulted in a much greater acceptance of numerical control in industry. This is because, unlike the N/C versions of the 1950s and 1960s, CNC today (see figure 1-8) does not require significant training. CNC machines have become easier to use with menu-selectable displays, advanced graphics, and conversational English programming. What used to be reserved for master-level machinists has now advanced to a point where entry-level machinists can rapidly learn and understand the technology. A general comparison of the characteristics of N/C versus CNC is given in table 1-1.

Fortunately, a lot of the work needed to program the tremendous variety of modern machines and controls has been simplified, much to the credit

TABLE 1-1 COMPARISON OF N/C AND CNC

N/C	CNC
Hard-wired	Software logic based
Control logic fixed (codes, functions, commands, etc.)	Control logic based in resident microprocessor
Internal control logic not changeable except through circuit board changes	Executive program changeable; makes control "think" like a machining center or turning center
No memory	Memory capacity with computational and compensational ability
Must be externally programmed and tape punched; no program created at machine	Program entered via direct line from external computer, floppy disk, cassette, punched tape, or manual data input
Programs stored on punched tape or cards	Programs stored on external computer and downloaded on demand via direct line; also stored in memory or on floppy disk or cassette
Tape must be recycled for every part	Memory capacity holds program; program executed from memory

FIGURE 1-8
A modern CNC unit.
(Courtesy of Cincinnati
Milacron Inc.)

of the manufacturers involved. The perforated or punched tape is rarely used today as an input medium. Direct and distributed numerical control, cassette tapes, and floppy disks are more prevalent throughout industry today, along with on-machine operator programming of the CNC unit itself.

ACCURACY, REPEATABILITY, AND RELIABILITY

Accuracy is defined as a measure of the extent to which a part is free from error. A machinist may be able to produce a part that is free from error. Many machinists can work to a tolerance of .001 in. on an old, worn machine if they are very familiar with that machine. A tremendous amount of experience and time is needed to be able to work consistently to this degree of accuracy. Some N/C machines today can produce accurate workpieces with tolerances of ±0.0002 in. or even ±0.0001 in., and the operator may not need to be familiar with the individual machine since the accuracy is built into the control and machine tool. These degrees of accuracy are not uncommon; some N/C machines can be purchased with systems capable of controlling much closer tolerances.

There are many factors that can affect a machine's accuracy. First, the foundation must be solid and must conform to the manufacturer's specifications. Second, the lubrication schedule and all proper maintenance procedures must be strictly adhered to at all times. In addition, cutting loads and forces, temperature of the environment, material to be machined, types of cutting tools, and toolholders can greatly affect the accuracy of any N/C or CNC machine.

Repeatability is a measure of the differences among the same dimensions of each piece machined. It is a function of the machine's ability to locate a machine slide within a given tolerance time after time. Slide repeatability to 0.000060 in. is becoming commonplace but is not always achieved in the cutting process. Spindle and tool drifts of 0.001 in. and greater often negate the benefit of precision slide systems.

The repeatability of N/C is roughly one-half of the actual positioning tolerance. The greater the accuracy and repeatability of the machine, the higher the cost. Repeatability is similar to accuracy in that the machine must receive proper care to maintain a particular level. The other factors affecting accuracy will also affect repeatability. Another element that should be pointed out is the care the operator must exercise in locating the parts in fixtures, vises, or whatever means are used to locate and clamp the workpiece. N/C operators should be aware that parts must be accurately located in the work-holding device, against positive stops. When the part is clamped, it is important to make sure it has not moved or been distorted out of position as a result of clamping forces. These simple checks ensure greater repeatability and quality of parts produced.

Reliability is another important goal of modern CNC manufacturers and users. Essentially it is a measure of dependability of the extent to which the machine can be consistently relied upon. Both control and machine tool reliability have increased considerably over the years. However, customers continue to demand greater accuracy and reliability of products. In order to meet and surpass this challenge, new types of slides, machine tables, bearings, lead screws, and advanced electronics are constantly being tried and tested.

The quality of a machine's parts and the tolerances with which they are manufactured and assembled are also important to machine reliability. Machine tool users, faced with rising costs, profitability pressures, and increasing competition, must have a product they can depend on to produce accurate parts. The machine tools should have few maintenance or downtime problems. For this reason, reliability will continue to be of great importance to machine tool builders and control manufacturers in future years.

REVIEW QUESTIONS

1. John Parsons is often considered the father of N/C. What educational institution helped refine early N/C?

2. What conditions in the United States accelerated the development of N/C?

3. Why is it necessary to design CNC machines and controls to be more user friendly and easier to operate?

4. Briefly explain the evolution of the N/C industry.

5. List and explain the principal differences between N/C and CNC.

6. Name different types of N/C machines. Which types are most common today?

7. To what degrees of accuracy are N/C machines able to produce?

8. Define repeatability. What is considered the general rule of thumb for determining the repeatability tolerances of an N/C machine?

9. What is the difference between accuracy and repeatability?

CHAPTER 2

What Is Numerical Control Programming?

OBJECTIVES

After studying this chapter, you will be able to:

- Discuss the importance of numerical control to manufacturing and productivity in the United States
- Define numerical control
- Explain the advantages and disadvantages of numerical control
- List the fundamental steps in planning for the use of N/C and CNC
- Describe general considerations and factors involved in N/C justification

IMPORTANCE OF NUMERICAL CONTROL

In the early 1940s, manufacturing in this country was heading in one direction—mass production. The economy was demanding a large volume of goods at competitive prices. *Automation* was the key to satisfying the demands of the market. However, automatic machines were expensive, and large lot sizes were required for these machines to be justifiable. At that time, mass production was aimed at the median of consumer taste. This resulted in mass mediocrity and, consequently, consumer dissatisfaction.

In the 1960s, the market, coupled with numerical control and other new production technologies, turned around completely. People began demanding a variety of products from which to choose. This required more versatile production equipment and smaller production runs. Manufacturers were concerned because universal machines were now needed where highly specialized machines were previously used. These conditions placed a great deal of responsibility on numerical control technology and the promise it held for maximizing profits and increasing productivity.

So what implication does that have for the importance of numerical control? Numerical control is more important now than ever. N/C has been the major force driving the metalworking industries to improved productivity levels. Closely coupled with the booming technologies of microelectronics and computers, it has truly revolutionized manufacturing.

The real importance of numerical control lies in the effects it has produced in this country. N/C machines are faster, more accurate, and more versatile where complex shapes are to be machined and where manual operations would otherwise be required.

Numerical control has risen in popularity through its ability to manufacture products of consistent quality more economically than alternative methods. Nevertheless, it is a popular misconception that numerical control is justifiable only for large-quantity production; just the opposite is true. An actual comparison of numerical control to standard machining methods indicates that the break-even point comes earlier with N/C production than with conventional production.

NUMERICAL CONTROL: WHAT IT CONSISTS OF

What exactly is numerical control, and what does it mean? Basically, it is control of machine tools by numbers, letters of the alphabet, and other symbols. These are translated into pulses of electrical current or other output signals that activate motors, drive systems, and other devices to run the machine tool. This collection of commands, logically organized into a "program," directs a machine tool in the machining of a workpiece.

Numerical control consists of three main elements: (1) the machine tool, which actually drives the cutting tools and performs the machining operations; (2) the control unit, which directs the positioning of the machine tool and guides the cutting tool actions; and (3) the program, which is the planned sequence of events feeding and directing the control unit for the specific machining actions required.

Three important points should be made about N/C. First, a machine tool operating under N/C can do nothing more than it was capable of doing before the control unit was joined to it. There are no new metal-removing principles involved. N/C machines position and drive the cutting tools, but the same milling cutters, drills, taps, and other tools still perform the cutting operations. Cutting speeds, feeds, and tooling principles must still be adhered to. Given this knowledge, what is the real advantage of numerical control? Primarily, the *idle time,* or time to move into position for new cuts, is limited only by the machine's capacity to respond. Because the machine receives *commands* from the machine control unit (MCU), it responds without hesitation. The actual utilization rate, or chipmaking rate, is therefore much higher than on a manually operated machine.

Second, numerical control machines can initiate nothing on their own. The machine accepts and responds to commands from the control unit. The control unit cannot think, judge, or reason. Without some input medium (e.g., punched tape, manual input, or direct computer link), the machine and control unit will do nothing. The N/C machine will perform only when the N/C data is prepared and loaded and the cycle start is initiated.

Third, the MCU can understand and act upon only its own language (the limited symbolic codes and functions that have specific meanings for that specific machine and control type). The programmer must know and understand these codes, commands, and functions and their precise meanings in order to input the information to the MCU and direct the actions of the machine tool.

ADVANTAGES AND DISADVANTAGES OF N/C

To begin a discussion of the advantages of numerical control, one must first realize that the average part spends only 5% of its manufacturing time on a machine in the shop. Only $1\frac{1}{2}$% of a part's manufacturing time is spent actually cutting metal, as shown in figure 2-1. N/C actually works to reduce the *nonchipmaking* time. With N/C performing such manual functions as selecting spindle speeds, feed rates, coolant control, and automatic fixture indexing, it has become generally accepted that N/C machines represent the most effective developments in manufacturing for reducing the unit cost of production.

As global competition intensifies and productivity and profitability pressures increase, management and labor become greatly affected by numerical control and its manufacturing capabilities. Some basic advantages of N/C have been mentioned already. However, a better understanding will come from the following explanations.

There are four basic phases that occur in most manufacturing. The first phase is *engineering,* or the determination of the product's size, shape, tolerances, and material. The second phase is *process planning,* which includes the decisions made concerning the selection of the manufacturing processes from order of operations to inspection standards. The third phase is *economic planning,* which includes determining economic lot sizes, raw materials, and inventory analysis. The fourth phase is *production,* including training of machine operators, machine setup, and actual machine operations.

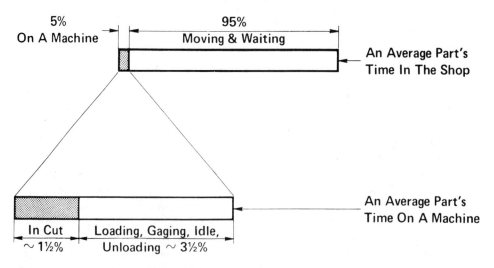

FIGURE 2-1
Breakdown of the time spent by an average part in the shop.

In conventional manufacturing, the production phase was frequently the only phase considered when judging the advantages of new developments in metalworking. All phases must be considered when judging N/C machines. N/C tends to reduce the importance of the production phase in relation to the others. When skillfully employed, N/C provides cost savings throughout the entire manufacturing process.

Before numerical control capability existed, engineers were severely limited in designing shapes using conventional machining. N/C makes it possible to produce even the most complex shapes without extremely high costs. Another advantage of N/C is the ability to make changes or improvements with a minimum of delay and expense. With conventional machines, it is often economically undesirable to make changes after the tooling is prepared. In addition, costs associated with conventional machines increase as tolerances become tighter. This factor caused engineering problems in trying to create parts with tolerances as loose as possible yet still capable of functioning properly. With N/C, tolerances are somewhat independent of costs. The machine always produces parts to maximize accuracy without special treatment. (This is true if the operator locates parts properly and pays attention to clamping forces, tooling considerations, and so on.)

As discussed in Chapter 1, N/C machines provide good positional accuracy and repeatability. Complex jigs and fixtures are not required in all cases. For most operations, the simplest form of clamping device is adequate. In addition to reducing complex fixtures, it is possible to reduce the use of expensive tooling. This factor greatly reduces the lead time required to get a new job into production.

Time study, in the conventional sense, is eliminated because the programmer now dictates how the part will be produced and how long it will take. After the program is established, there can be no variations from part to part and no deviation from the programmed time.

A high degree of quality is inherent in the N/C process because of its accuracy, repeatability, and freedom from operator-introduced variations. In-process quality inspection is seldom required after an inspection of the first part produced from new N/C data as a check on the programming function. A coordinate measuring machine (see figure 2-2) is used to check positional accuracy.

One of the basic functions of economic planning, as previously suggested, is the determination of economic lot size. With conventional machining methods, setup costs are high and cannot be calculated with any degree of accuracy. Therefore, it is necessary to make a large number of parts for each setup if the unit part cost is to be minimized. With numerically controlled machines, the high process predictability ensures accurate cost determinations, and the simplified, low-cost setups enable parts to be run in small quantities economically.

In addition, since the programmer selects the methods and the sequence of operations, as well as operating feeds and speeds, cutting conditions are under the complete control of manufacturing supervision. With N/C, actual physical manipulation of the machine by the operator is greatly reduced because feeds and speeds are, in most cases, automatically selected.

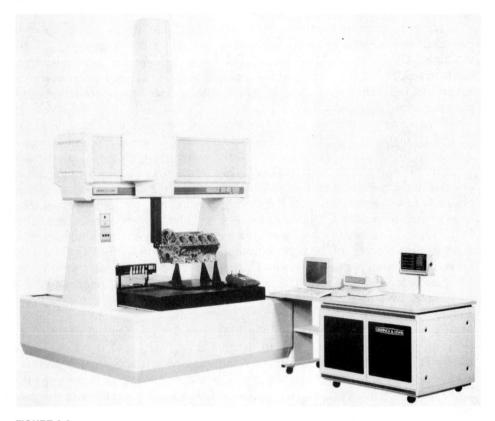

FIGURE 2-2
A coordinate measuring inspection machine. (Courtesy of Giddings and Lewis Measurement Systems)

Some other advantages of numerical control are as follows:

- *Reduced scrap and rework.* Errors due to operator fatigue, interruptions, and other factors are less likely to occur on N/C machines.

- *Improved production planning.* N/C machines can often perform, at one setting, work that would normally require several conventional machines.

- *Reduced space requirements.* Since fewer jigs and fixtures are used, the actual storage requirements of these expensive tools are reduced.

- *Simplified inspection.* Once the first piece has passed inspection, minimal inspection is required on subsequent parts.

- *Lower tooling costs.* There is less need for complex jigs and fixtures.

- *Reduced lead time.* This is a result of lower tooling costs.

- *Greater ease in accomplishing complex machining operations.* This is due to advanced machine control and programming capabilities.

- *Increased machine utilization.* This means that the N/C machine tool runs more and stays "in the cut," resulting in less idle time.

- *Increased part consistency and uniformity.* This is a result of less dependency on an operator.

- *Reduced parts inventory.* With N/C and CNC, inventory is reduced because small lot sizes can be run quickly with less advance notice and lead time.

There are, however, some factors relating to numerical control that some individuals might call disadvantages or inhibitors to using numerically controlled equipment. Some of these disadvantages are worth mentioning, but a detailed analysis will in most cases reveal that the advantages of numerical control outweigh the disadvantages.

First, tools on N/C machines will not cut metal any faster than tools on conventional machines. N/C machines merely position and drive the cutting tools. Optimized feeds and speeds can be run on either conventional or N/C machines.

N/C does not eliminate the need for expensive tools. Some jobs require special and expensive fixtures and cutting tools. The most significant factor is the greater initial cost of the N/C machine relative to that of a conventional machine. Machines and tooling are costly today, and their purchase requires extensive justification.

Another factor that must be considered is that, contrary to popular belief, N/C will not totally eliminate errors. Operators can still push the wrong buttons, make incorrect alignments, and fail to locate parts properly in a fixture. Some of these types of errors can be minimized by careful and effective training. However, some errors will always be likely to occur; they will never be totally eliminated.

Proper selection and training of programmers and maintenance personnel is required. The support personnel are essential to the success of any N/C installation and must be given careful and adequate consideration.

These factors may not be considered disadvantages of N/C so much as inhibitors against purchase. Undoubtedly, some smaller companies have decided not to purchase N/C equipment after weighing all costs and requirements involved. As with any advanced technological equipment, N/C should be used only where it will produce the work better, faster, and more accurately than conventional methods. Many shops, though, after reluctantly purchasing their first piece of N/C or CNC equipment, have found the actual savings and advantages to be much greater than originally planned.

PLANNING FOR THE USE OF N/C

What are the actual steps a potential N/C user takes prior to purchasing N/C equipment? There are 10 factors that should be considered before using numerically controlled equipment. Depending on the requirements of the facility, equipment to be purchased, and type of work to be produced, additional consideration may be necessary.

Basic Knowledge of How N/C and CNC Works

For any prospective user, there is no substitute for firsthand information concerning CNC machine and control unit operations and capabilities. One of the best ways in which potential CNC users can obtain this information is by attending CNC training programs and demonstrations offered by machine tool manufacturers and suppliers.

Capital Investment Requirements

As previously discussed, CNC machines require a greater initial investment in comparison with standard machines. In numerical control, as applied to basic machine tools, the basic machine must be designed with a more rigid frame and heavier lead screws, bearings, and other actuating mechanisms. These allow the machine tool to achieve the acceleration speeds required for efficient positioning times. Added to the initial cost of the machine are N/C data preparation and processing expenditures, as well as special tooling and other costs necessary to maintain an N/C environment. The net capital expenditure may be in the range of $30,000 to $300,000 for a CNC lathe or machining center to more than $2 million to $3 million for highly sophisticated five-axis profilers. A careful survey will reveal that despite these high initial costs in some cases, a properly operated N/C installation will pay for itself in a remarkably short period of time.

Personnel and Training

One of the most important points to keep in mind when considering the purchase of CNC machinery is that, in order to get the maximum return from the capital invested, good cooperation and communication must be maintained among shop, programming, and engineering personnel. N/C opens up great opportunities for the engineer and designer to use design techniques that were impossible when only conventional manufacturing methods were available.

When selecting personnel for operator, programmer, and maintenance positions, each company should survey its current employees for essential skills that these high-technology jobs require. It may be necessary for a company to hire outside its establishment rather than train existing personnel in order to obtain the required skills. Selecting qualified and skilled personnel is critical to the success of a CNC installation. This selection process must be approached cautiously and objectively; considerable thought, discussion, and research must be conducted prior to the installation of equipment.

Programming and Tooling

One aspect in any consideration of CNC equipment is how to handle programming requirements. Later chapters will discuss the CNC preparation and communication equipment necessary to support the programming function and whether manual or computer methods of programming should be used.

Tooling for N/C is closely related to programming because it involves the choice, size, and shape of cutters to match the plan visualized by the programmer. It is customary for the program to include tool specifications so that the machine operator can get the complete set of tools for a given part without initially having to work out a requirements list. The type of CNC system under consideration will determine whether the programmer will use standard tools from a tool store or whether an elaborate tool library must be established for preset tools.

Maintenance and Repair

Another concern for management is maintenance. Some companies have been discouraged from installing N/C and CNC machines because they felt that they did not have adequate facilities or personnel to service the electronically controlled equipment. With the advent of integrated circuits, microprocessors, and computer numerical control, there has been a general increase in the reliability of controls. Still, the lack of trained or capable personnel is a concern for potential CNC users. However, maintenance classes are offered by control manufacturers to educate new maintenance personnel and update current personnel. It should be noted that maintenance, as well as programming and operation of the equipment, must be restricted to trained and authorized personnel for purposes of safety and equipment protection.

Cost Analysis

Generally, in a discussion of the cost analysis of numerical control, the programming costs and N/C data preparation time are being compared with the design and upkeep of the jigs, fixtures, and setups required for conventional machining operations. The elimination of elaborate jigs and fixtures probably constitutes the largest area of savings in numerical control.

Another item of importance in cost analysis is the shorter lead time of CNC machined parts. Again, a typical part will spend about 5% of its manufacturing life being machined and the other 95% sitting in flats, being inspected, and waiting. With N/C, parts are routed to fewer machines, thus cutting down the total manufacturing time.

Quality Control

One benefit of numerical control is the repeatability of parts produced and the reduced inspection time. N/C has made it possible to produce part after part with consistent accuracy. The adverse effects of operator skill, fatigue, and human reliability have been reduced to a minimum. More complex parts can now be produced with much lower rejection rates than with conventional methods.

Reduction of Inventory

With the advent of N/C and CNC, shorter lead times are needed, thereby reducing the total amount of inventory required. Finished inventory can be held to an absolute minimum because of the care and repeatability with which a part can be put in process and the resulting speed with which inventory can be replaced. Raw material inventories can also be cut because it is no longer necessary to schedule long runs.

Environmental Requirements

One of the most important factors to be considered in studying a plant layout for an N/C installation is accessibility. The flow of work to and from the machines is important because of the tremendous "appetite" of N/C and CNC machines. All plant layouts should be based on the maximum use factor of these units. Other environmental considerations should include cleanliness and possible extensive chilling or heating conditions.

Service Responsibility

Because greater emphasis is placed on numerical control systems for manufacturing, it is essential that these systems be kept in operation as much as possible. It is often difficult to determine whether a problem has originated with the control or the machine tool. This sometimes results in a situation where the user does not know whether to contact the machine tool service representative or the control service representative. Consequently, much time can be wasted trying to determine the source of the problem. Many companies will purchase both machine and control from the same manufacturer, thereby eliminating the question of who will make the repair.

N/C JUSTIFICATION

N/C justification has already been discussed in relation to the advantages of numerical control and planning for its use. However, some additional concepts should be examined.

At present there are several approaches to equipment justification. One method is the cost-savings approach. It has the distinct advantage of being easy to calculate. It is basically conservative, based on equipment replacement with some degree of improved productivity and performance. The disadvantage of this approach is that it involves no disciplined effort to review the entire operation for improvement.

Another widely used justification technique is aggressive justification. This method turns the cost-savings approach disadvantage into an advantage. Aggressive justification questions whether the present methods are the best; it may involve substantial changes in manufacturing methodology. There are tremendous opportunities for processing improvements with this method, but the translation into dollars and cents is complex.

Whenever an N/C justification is needed, an analysis of the parts to be programmed is an ideal place to start. Once it is determined that a realistic N/C work load exists, the second step is to determine the return on investment (ROI). For the calculation of the ROI, it is necessary to study the different aspects of the business that are affected by the use of this new equipment, not just the machine itself. One of the questions that should be answered is "Will N/C help produce a better- and consistent-quality part in an economical manner?" The calculation is computed by dividing the average savings per year by the initial investment cost. The result is a rate of return.

$$\frac{\text{Average savings/year}}{\text{Investment}} = \text{Rate of return}$$

Perhaps some productivity comparisons will also have to be made based on estimated production lot size (see figures 2-3 and 2-4).

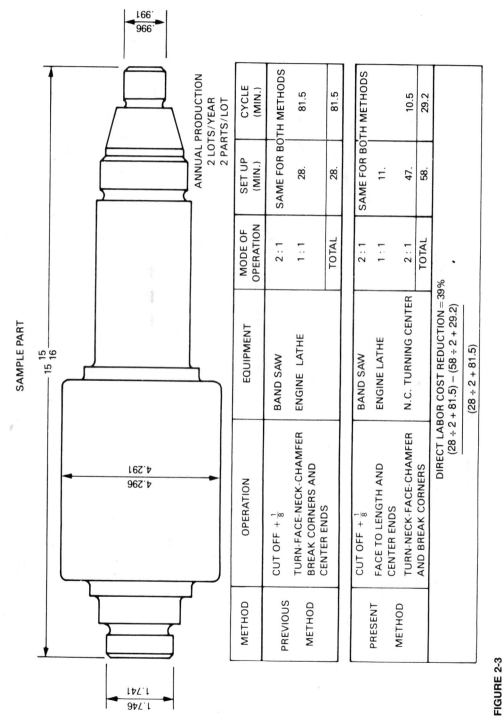

SAMPLE PART

ANNUAL PRODUCTION
2 LOTS/YEAR
2 PARTS/LOT

METHOD	OPERATION	EQUIPMENT	MODE OF OPERATION	SET UP (MIN.)	CYCLE (MIN.)
PREVIOUS METHOD	CUT OFF + 1/8	BAND SAW	2 : 1	SAME FOR BOTH METHODS	
	TURN-FACE-NECK-CHAMFER BREAK CORNERS AND CENTER ENDS	ENGINE LATHE	1 : 1	28.	81.5
			TOTAL	28.	81.5
PRESENT METHOD	CUT OFF + 1/8	BAND SAW	2 : 1	SAME FOR BOTH METHODS	
	FACE TO LENGTH AND CENTER ENDS	ENGINE LATHE	1 : 1	11.	
	TURN-NECK-FACE-CHAMFER AND BREAK CORNERS	N.C. TURNING CENTER	2 : 1	47.	10.5
			TOTAL	58.	29.2

DIRECT LABOR COST REDUCTION = 39%

$$\frac{(28 \div 2 + 81.5) - (58 \div 2 + 29.2)}{(28 \div 2 + 81.5)}$$

FIGURE 2-3
Productivity comparison: small lot sizes.

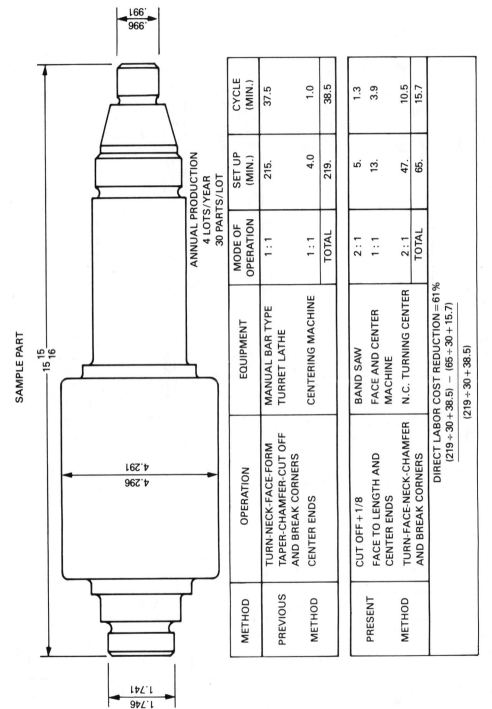

SAMPLE PART

.996
.991

15 15/16

4.296
4.291

1.746
1.741

ANNUAL PRODUCTION
4 LOTS/YEAR
30 PARTS/LOT

METHOD	OPERATION	EQUIPMENT	MODE OF OPERATION	SET UP (MIN.)	CYCLE (MIN.)
PREVIOUS METHOD	TURN-NECK-FACE-FORM TAPER-CHAMFER-CUT OFF AND BREAK CORNERS	MANUAL BAR TYPE TURRET LATHE	1 : 1	215.	37.5
	CENTER ENDS	CENTERING MACHINE	1 : 1	4.0	1.0
			TOTAL	219.	38.5

METHOD	OPERATION	EQUIPMENT	MODE OF OPERATION	SET UP (MIN.)	CYCLE (MIN.)
PRESENT METHOD	CUT OFF + 1/8	BAND SAW	2 : 1	5.	1.3
	FACE TO LENGTH AND CENTER ENDS	FACE AND CENTER MACHINE	1 : 1	13.	3.9
	TURN-FACE-NECK-CHAMFER AND BREAK CORNERS	N.C. TURNING CENTER	2 : 1	47.	10.5
			TOTAL	65.	15.7

DIRECT LABOR COST REDUCTION = 61%

$$\frac{(219 \div 30 + 38.5) - (65 \div 30 + 15.7)}{(219 \div 30 + 38.5)}$$

FIGURE 2-4
Productivity comparison: production lot sizes.

Regardless of the method of N/C justification that is used, much more detailed work and analysis are needed beyond what is mentioned in this chapter. These are merely two of the many different types of N/C justification techniques.

REVIEW QUESTIONS

1. What were some of the important factors leading to the popularity of numerical control?

2. In your own words, briefly define numerical control.

3. The average part spends only 5% of its manufacturing time on a machine. What comprises the other 95% of an average part's time on the shop floor?

4. What are the four basic phases that occur in most manufacturing? Briefly explain their importance and relationship to one another.

5. List and briefly describe the major advantages of numerical control. What are the disadvantages of or inhibitors to numerical control?

6. What is the major advantage of numerical control over conventional machining?

7. List and briefly describe the 10 basic considerations in planning for the use of N/C.

8. Name two techniques used for successful N/C justification, and briefly explain their functions. Which approach is most likely to yield the greatest overall savings? Why?

CHAPTER 3

How N/C and CNC Operate

OBJECTIVES

After studying this chapter, you will be able to:

- Explain how a workpiece is processed for numerical control
- Discuss how N/C and CNC collect, store, and act on data input
- Identify the primary types of numerical control systems
- Discuss the common types of N/C drive and feedback systems
- Explain coordinate systems and machine axes
- Discuss absolute and incremental dimensioning and control systems
- Describe the functionality of zero shift and offset

WHAT A MACHINIST NEEDS TO KNOW

Since the number of skilled machinists in this country is on the decline, the skills themselves must be compensated for in another manner. The machine and control unit, through the direction of the programmer, now perform many of the functions of the skilled machinist.

The entire process of N/C programming depends on the programmer's ability to visualize the actual cutting motions and table movements that are taking place on the machine. These movements are then translated into a coded format. Consequently, many interactive decisions and judgments are made by programmers as well as machinists.

Let us study some of the judgments and functions of a machinist. A machinist will begin a workpiece with a thorough study of a blueprint, sketch, or sample workpiece. In addition, a machinist will check the process routing sheet, if available, to determine the specific machining to be performed for that particular operation. If specific tolerance requirements are needed, such as an allowance for grind stock, the process sheet should contain this information for the machinist's review prior to machining. Another important step in machining a workpiece is planning the sequence of machining operations. The concept of "What do I do first; what do I do next?", along with determining how the individual setup is to be made, is of prime importance. These decisions are closely related to the experience level of the machinist. A machinist also calculates the speeds and feeds, and selects cutting tools, materials, and machine tools where appropriate.

There may be more detailed items that an individual machinist must be concerned with, but essentially the machinist *visualizes a program*. This program tells what workpiece is to be machined and how. The machinist then guides the cutting tools in their relationship to the workpiece by means of the operating dials and levers. After the machining operations are complete and the material is removed from the workpiece, an accurate part is produced. In essence, a skilled machinist has programmed, accumulated, stored, and transmitted detailed information in order to produce a particular workpiece.

WHAT A PROGRAMMER NEEDS TO KNOW

A *programmer* must make use of all the skills of a machinist in processing a part. Only recently has the effective operation of the machine tool become the programmer's responsibility. A programmer's sequence of machining operations; selection of cutting tools, feeds, and speeds; and basic programming methodology are critical to the efficient use of the machine used to process the part. In addition to visualizing the entire process, the programmer must also be familiar with the particular machine being programmed and its general operating characteristics and requirements. The programmer must also understand the specific N/C or CNC language or format that the particular machine control unit will accept. An understanding of computer programming languages may be necessary to obtain the desired output results. What the operator has traditionally done with intuitive insight, training, and years of experience, the part programmer must now do with intellect and reasoning combined with practical know-how. The result then is the completed part program for a specific part. A flow diagram of the steps in processing an N/C program is illustrated in figure 3-1.

The process of becoming an N/C programmer involves some critical elements. Principal among these is skill in print reading. The programmer must be able to visualize, in three dimensions, parts depicted on a two-dimensional part print. Knowledge of metal-cutting processes and cutting tools is also important. Another desirable quality, but one that is becoming increasingly difficult to obtain, is experience on the shop floor solving the multitude of manufacturing problems. This type of individual represents the optimum mix.

Actually, N/C programmers can vary from being coders to being comprehensive manufacturing engineers responsible for planning the entire sequence of machining operations and selecting the appropriate tools and fixtures as well as preparing the N/C coding. These individuals will continue to be in demand as companies strive to become more competitive and to improve the quality and productivity of their products and operations.

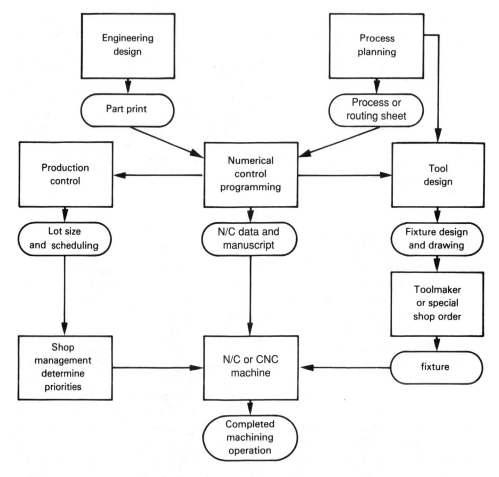

FIGURE 3-1
A flow diagram of the steps in processing an N/C program.

MACHINE REGISTERS AND BUFFER STORAGE

N/C data enters the machine control unit through one of five means (to be discussed in detail in Chapter 4). These include (1) tape readers, (2) cassette tapes, (3) floppy disks, (4) direct lines to computers (known as DNC), and (5) manual data input (MDI).

The N/C data, or coded information, is then passed on to *registers* within the control. These registers accept the information, which consists of machine coordinates, preparatory functions, and miscellaneous functions. This information is then transmitted to the respective register sections, where actuation signals are relayed to the machine tool drives. A basic sketch of this process is illustrated in figure 3-2.

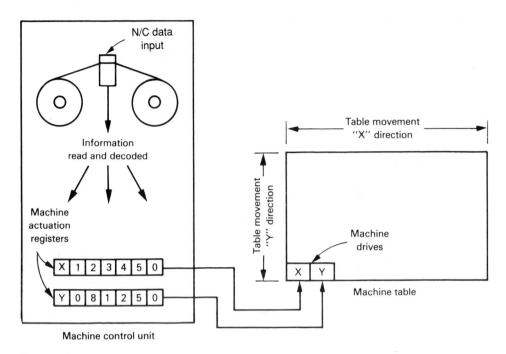

FIGURE 3-2

Basic sketch of N/C data input information being read, decoded, and passed to machine actuation registers, resulting in corresponding table movements.

Most modern N/C and CNC controls are equipped with *buffer storage.* As shown in figure 3-3, this feature allows the control to accept information into a buffer register while an operation is being performed from the active machine registers. When that operation is completed, the information is transferred from buffer storage to the machine actuation registers. This transfer of information is instantaneous, thereby reducing the time between N/C data input, such as tape reading, and machine performance. Buffer storage reduces the amount of *dwell time* between machine operations because the next N/C instruction is read and stored while the machine is executing the previous block. Part finish is also better because the machine does not have to wait for the new N/C data to be read and interpreted before entering the machine actuation registers. Such hesitation could cause the cutter to dwell, thereby marking the part surface.

AXIS RELATIONSHIPS

Once the N/C data has entered the MCU and the machine actuation registers are loaded, the machine responds with its appropriate coordinate-axis movements and other commands. The primary axis movements to be studied in terms of their relationship to each other are the X- and Y-axes. *Axis* refers to any direction of motion that is totally controlled by specific N/C commands.

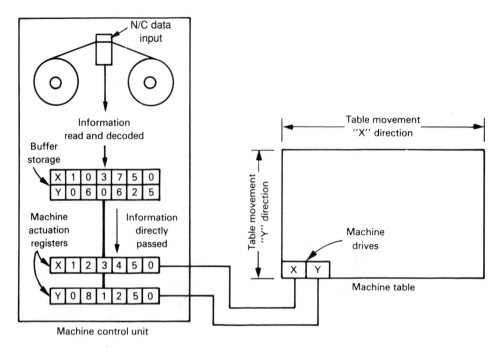

FIGURE 3-3

Basic sketch of N/C data input information being read, decoded, and stored in buffer storage until machine actuation registers have completed previous move and commands. Information is then passed from buffer storage to actuation registers.

Machines with only X and Y positioning capability are known as two-axis machines. On machines of this type, distances in the Z direction are controlled by the operator or by preset stops similar to that of a conventional drill press. An example of table movements in the XY plane is shown in figure 3-4. On other N/C machines, the Z depth is programmable, thereby making the machine a three-axis machine (X, Y, and Z). The three basic motions, as designated by the Electronic Industries Association (EIA), are X, Y, and Z. The principal X motion is parallel to the longest dimension of the primary machine table. The primary Y movement is normally parallel to the shortest dimension of the primary machine table. The Z motion is the movement that advances or retracts the spindle. The Z-axis movement changes by the fact that N/C machines are made with vertical and horizontal spindles. To help understand the Z-axis, it can be said that a line through the center of the machine spindle is actually the Z-axis. It is only when the actual depth of cut (Z-axis) is programmatically controlled that the machine is considered a true *three-axis* N/C machine. That is, the machine is capable of *simultaneous* motion in the X, Y, and Z directions. Additional machine/axis relationships can be seen in figures 3-5, 3-6, 3-7, and 3-8.

As the N/C machine positions itself corresponding to the programmed positions input to the MCU, the positions obtained are displayed on the CNC's

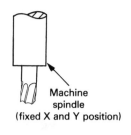

Machine
spindle
(fixed X and Y position)

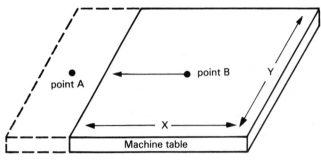

point A   point B   Y

X

Machine table

FIGURE 3-4
A machine table
movement to the left
in the X direction is
needed to move from
point A to point B.

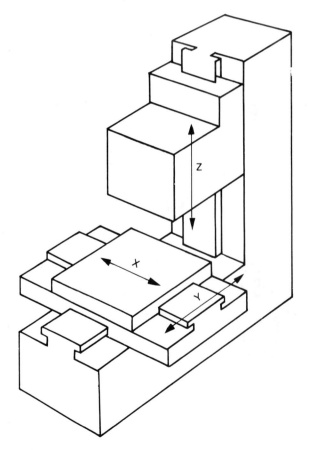

Z

X

Y

FIGURE 3-5
A vertical N/C machine.

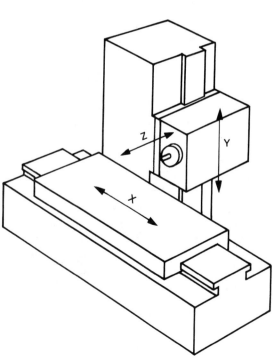

Z   Y

X

FIGURE 3-6
A horizontal N/C machine.

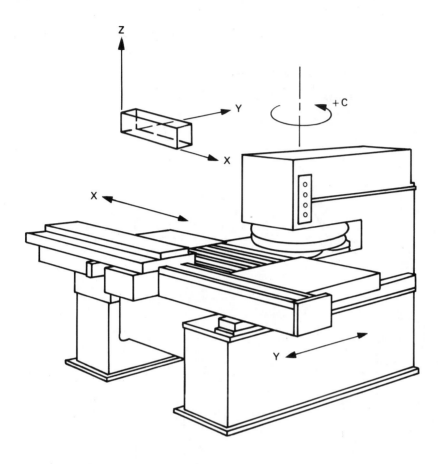

FIGURE 3-7
An N/C turret
punch press.

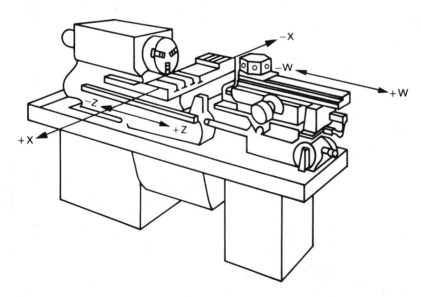

FIGURE 3-8
An N/C turret
lathe.

FIGURE 3-9
A typical CNC unit with CRT screen that displays X, Y, and Z command
positions, spindle speeds, feed rate, tool number, and other information.
(Courtesy of Cincinnati Milacron Inc.)

display screen (the cathode ray tube, or CRT screen). The readouts are generally
for *sequence number; X, Y, and Z command positions; spindle speed; feed rate; tool
number;* and other information. An example of a CNC CRT screen is shown in
figure 3-9.

The individual sequence numbers identify the block or line of information
being read. The X, Y, and Z position readouts are constantly changing as the
program advances and are mainly provided so that the operator can identify
specific positions and lines or blocks of information relative to the program
manuscript.

DRIVE MOTORS AND MOTION

Machine tool drive motors are of four types: stepper motors, DC servos, AC
servos, or hydraulic servos. Stepper motors move a set amount of rotation (a
step) every time the motor receives an electronic pulse. DC and AC servos are
widely used variable-speed motors found in small and medium continuous-
path machines. Unlike a stepper motor, a servo does not move a set distance;
when current is applied, the motor starts to turn, and when the current is removed, the motor stops turning. The AC servo is a fairly recent development.
It can develop more power than a DC servo and is commonly found on CNC

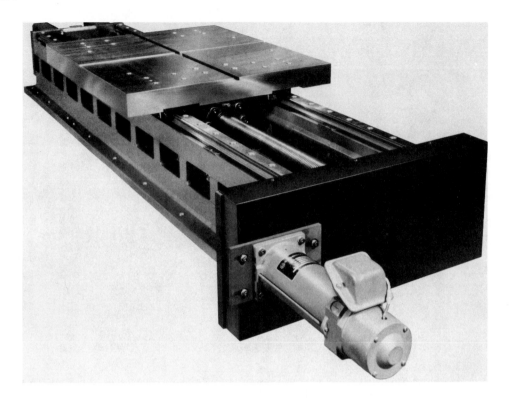

FIGURE 3-10
A CNCs electric servomotor. (Courtesy of Cincinnati Milacron Inc.)

machining centers. The typical electric servomotor system has the feedback system and motor in the same housing (see figure 3-10). Recent advances in the power and response of electric motors have made them the preferred choice for all CNC machines.

Using electric motors permits a reduction in the size of (or even total elimination of) the hydraulic system. Machines with electric drives may still use a small hydraulic system for tool clamping and changing.

Hydraulic servos, like AC or DC servos, are variable-speed motors. Because they are hydraulic motors, they are capable of producing much more power than an electrical motor and have been used on large N/C machinery.

The rotary motion from the drive motors is converted to linear motion by recirculating ball-lead screws called *ballscrews*. This type of screw (see figure 3-11) has low friction and almost no backlash and can run at very high speeds. The machine axes travel on low-friction roller bearings instead of traditional ways and gibs to allow for higher table speeds. Older machines without roller bearings are typically limited to less than 100 in./min. maximum table speed.

Electrically driven machines can have table travel speeds in excess of 400 in./min. and as high as 1200 in./min. or higher for certain types of CNC ma-

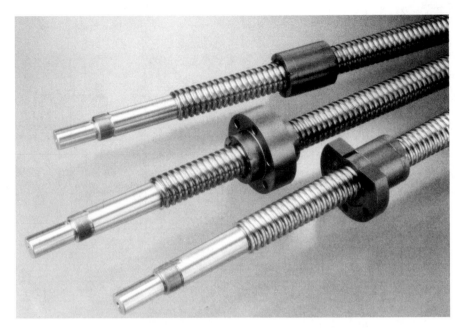

FIGURE 3-11
Recirculating ball-lead screws called ballscrews. (Courtesy of Thomson Industries)

chines. These high table speeds are used for positioning the machine between machining operations, but actual machining seldom takes place at speeds above 150 in./min.

FEEDBACK SYSTEMS

Once the command signals have been sent from the machine control unit to the machine, the slide motion and spindle movement occur. How then does the N/C control know that the machine is properly positioned? Unless the control and machine form a *closed-loop system*, the control really has no way of knowing if the machine is properly positioned. Figure 3-12 illustrates a basic closed-loop system. Such systems are similar in operation to driving an automobile. When driving an automobile, an individual will check the speedometer to determine speed. The actual speed is compared to the desired speed designated by the speed limit signs. The driver's brain detects the difference between the posted speed and the actual speed, and the brain then instructs the foot to adjust the car's pedals until the correct speed has been achieved.

Open-loop systems provide no check or measurement to indicate that a specific position has actually been achieved. No feedback information is passed from the machine back to the control. The system components may be affected by time, temperature, humidity, or lubrication, and the actual output may vary from the desired output. The main difference between open- and closed-loop systems is that with closed-loop systems, the actual output is measured, and

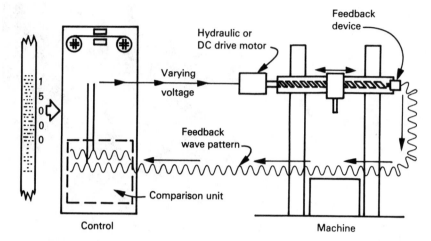

FIGURE 3-12
A closed-loop system.

a signal corresponding to this output is fed back to the input station, where it is compared with the input registers. Such a system automatically attempts to correct any discrepancy between desired and actual output.

Feedback systems may be either *digital* or *analog*. Digital systems generate pulses that are fed back to the control and count down linear or rotary motion in minimum movements on machine lead screws. Older analog systems sense and monitor variations in levels of voltage. Moving tables on a machine may overshoot in both directions and then search for the exact position to stop.

TYPES OF NUMERICAL CONTROL SYSTEMS

Older hard-wired numerical control systems were generally classified into two basic types: positioning, or point-to-point; and continuous-path, or contouring.

Positioning, or point-to-point, programming, as illustrated in figure 3-13, is best described as moving or directing a tool to a specific location on a workpiece to perform operations such as drilling, tapping, boring, reaming, and punching. The process of positioning from one coordinate (X, Y) position to another and performing these basic operations continues until all work has been performed for programmed locations. The important aspect of positioning systems is that, on a true positioning system, the cutting tool is not in constant contact with the workpiece. The spindle or the table may move to locate the desired position directly under the spindle. When the X and Y positions are satisfied, the spindle will then advance the cutting tool into the workpiece. There are some positioning systems that do possess limited contouring capabilities such as straight-line milling along either the X- or Y-axis. In addition, 45° angle milling is possible on some positioning systems with limited contouring capabilities.

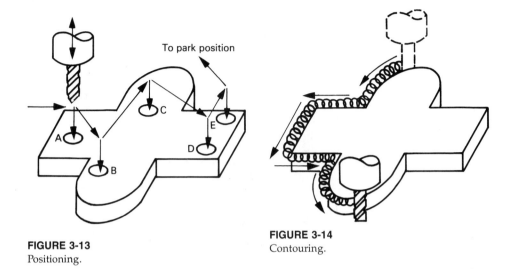

FIGURE 3-13
Positioning.

FIGURE 3-14
Contouring.

Contouring, or *continuous-path, systems* maintain a constant cutter-workpiece relationship. The cutting tool remains in constant contact with the workpiece as the corresponding coordinate movements are attained. This process is illustrated in figure 3-14. The most common of the continuous-path operations are milling and lathe operations that profile workpieces to exact specifications. The actual contouring control system must have speed control independent of the X and Y driving motors. This enables the rate of travel to be regulated on at least two axes at the same time.

The distinction between positioning and contouring control systems has been blurred over time because of the progress made in sophisticated electronics and the advent of CNC. In addition, workpiece complexity, tighter tolerances, and improved part finishes have also forced the transition to more capable systems. Virtually all new CNC units are capable of handling both positioning and contouring requirements.

CARTESIAN COORDINATE SYSTEM

The entire concept of numerical control is based on the principle of rectangular coordinates discovered by the French philosopher and mathematician René Descartes. This mathematical development, made over 300 years ago, is better known as the *Cartesian coordinate system.* Through the use of rectangular coordinates, any specific point in space can be described in mathematical terms along three perpendicular axes. The idea applies perfectly to machine tools because their construction is generally based upon two or three perpendicular axes of motion.

The system of Cartesian coordinates is illustrated in figure 3-15. As in algebra, the *X-axis* is horizontal (left and right), and the *Y-axis* is vertical (up and down). The *Z-axis* is then applied by holding a pencil perpendicular to the paper with its point at the location where the X and Y lines cross each other. The point where the X- and Y-axes cross is called the origin, or zero point. Four

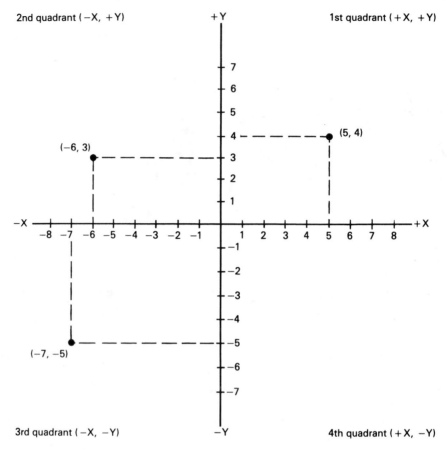

FIGURE 3-15
Cartesian coordinate system.

quadrants are formed when the X- and Y-axes cross. Each quadrant is numbered in counterclockwise rotation (see figure 3-15).

The plus (+) and minus (−) signs indicate a direction from the zero point along the X and Y areas. As seen from figure 3-15, all positions plotted in the first quadrant have positive X and positive Y values (+X,+Y). All positions in the second quadrant have negative X and positive Y values (−X,+Y). Points plotted in the third quadrant have values of negative X and negative Y (−X, −Y). In the fourth quadrant, all positions have values of positive X and negative Y (+X,−Y).

At first glance, it appears that it would be easier if all work could be done in the first quadrant since all values are positive and no signs would be needed. However, any of the four quadrants may be used on different machines. In some applications, the use of minus signs is a distinct advantage. Therefore, programmers must be thoroughly familiar with the use of both plus and minus signs in all four quadrants.

AXIS CONTROL

Two-axis tape control normally consists of the X- and Y-axes. In figure 3-15, it can be seen that the point labeled (5,4) actually means X = 5.0000 and Y = 4.0000 from the zero point. The point (−7, −5) means X = −7.0000 and Y = −5.0000 in relation to the zero point. It is generally agreed that in this kind of notation, the X dimension is written first, followed by the Y dimension—that is, (X, Y).

An older but simple two-axis N/C machine is shown in figure 3-16. In most cases, the machine table moves left and right to position the cutter. A longitudinal movement of the machine table on this machine is an X-axis move. A cross, or saddle, movement is a Y-axis move. In order to obtain identical X-Y movements on some two-axis machines, the table remains stationary and the spindle moves to satisfy X and Y locations. The movement of the cutter remains the same regardless of whether the spindle or table positions the workpiece. This consideration is taken care of by the machine and control manufacturers. The programmer must specify only the dimensions and the plus or minus signs

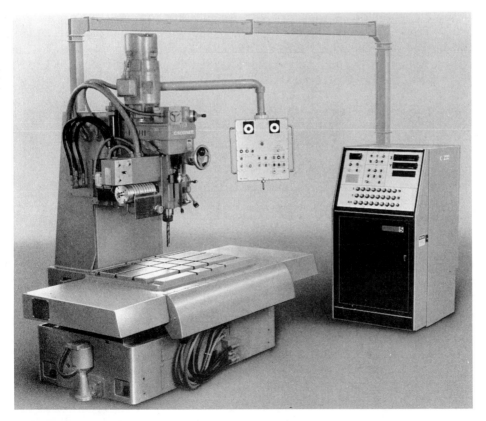

FIGURE 3-16
A typical two-axis N/C machine. (Courtesy of Cincinnati Milacron Inc.)

in relation to the zero point. How the coordinate system relates to the machine table will be explained in detail later.

In numerical control programming, plus signs may be printed beside all positive locations. A plus sign normally does not need to be printed if the value is positive unless it is required by that particular N/C machine and control type. Minus signs must be written to distinguish between negative and positive values when the sign has been omitted.

Z-axis movements, just like X and Y movements, are programmable. As explained earlier, a line through the center of the machine spindle is the Z-axis. Figure 3-17 illustrates a vertical Z-axis with the X- and Y-axes drawn in relation to the workpiece and machine table. A positive Z movement is described as a movement of the tool *away* from the work. A negative Z movement is described as a plunge cut, or a movement of the tool *into* the workpiece. Figure 3-18 illustrates a horizontal Z-axis, with the X- and Y-axes drawn in relation to the workpiece and machine table. Positive Z movement moves the tool *away* from the work, and negative Z movement moves the tool *into* the work. Notice that because the workpiece is positioned in relation to the zero point, all positions on the X-Y plane surface will have positive values and will be situated in the first quadrant.

Some of the more advanced CNC machines are equipped with four- and five-axis contouring capabilities. These capabilities generally involve secondary and tertiary axes that run parallel to the X-, Y-, and Z-axes, plus some addi-

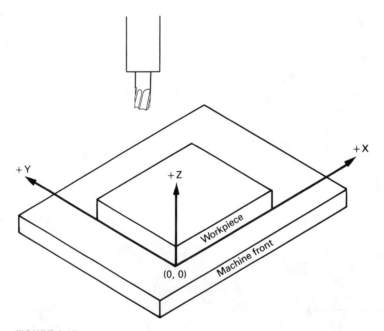

FIGURE 3-17
Vertical Z-axis drawn in relation to the X- and Y-axes, workpiece, and machine table.

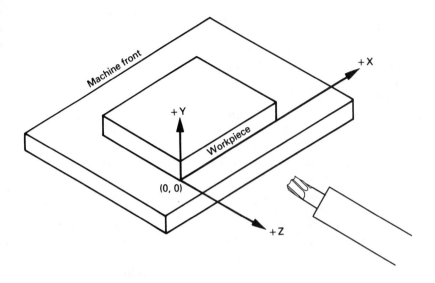

FIGURE 3-18
Horizontal Z-axis drawn in relation to the X- and Y-axes, workpiece, and machine table.

tional rotational movements within the coordinate locations. These rotational movements consist of three primary axes of motion: A, B, and C. Rotary axes of motion rotate around the three primary axes of motion X, Y, and Z. The A-axis rotates around a line parallel to the X-axis, B around Y, and C around Z, as shown in figure 3-19.

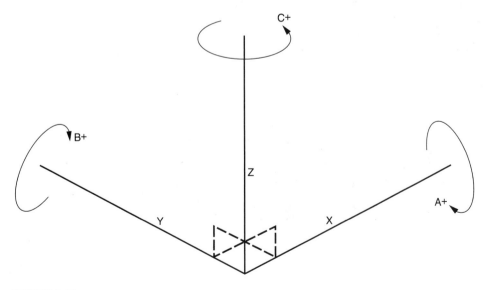

FIGURE 3-19
Rotary axis of motion: A-axis rotates around a line parallel to the X-axis, B around Y, and C around Z.

INCREMENTAL AND ABSOLUTE SYSTEMS

Prior to any discussion involving incremental programming systems, students must have a thorough understanding of incremental dimensioning. Carefully study the part in figure 3-20. The distance from the left edge of the part to hole 1 is 1.25. From hole 1 to hole 2, the distance is 1.50. The distance from hole 2 to hole 3 is 1.50 and from hole 3 to hole 4 is 1.62. This is known as *incremental* dimensioning. It is also referred to as delta dimensioning. The word "delta" is derived from a Greek letter used to denote the difference between two quantities. In figure 3-20, each dimension is given incrementally from the last position to the next position.

An *incremental system* works according to this principle; it positions the work or cutter in increments from the immediately preceding point. Calculations are made from the location of the tool or table to where it is going. The use of plus and minus signs involves a new aspect when used in the incremental mode. A positive X move does not refer to a specific rectangular quadrant, but directs the tool to move to the right along the X-axis from its current position. A negative X move directs the tool to the left. Similarly, a positive Y move positions the cutter up from the present location, and a negative Y is a command to move down. A positive Z directs the cutter away from the workpiece, and a negative Z is a move toward or into the workpiece.

A close examination of the workpiece in figure 3-21 will reveal its similarity to that shown in figure 3-20. The difference is the way the actual part is dimensioned. This type is known as *absolute*, or baseline, dimensioning because all positions must be given as distances from the same zero location or reference point. All dimensions are calculated from one zero point, as indicated in figure 3-21.

An *absolute system* operates according to absolute dimensioning. All programmed positions are figured from the same zero or reference point. All po-

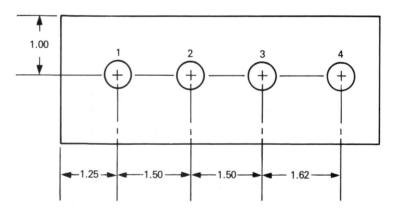

FIGURE 3-20
Incremental dimensioning.

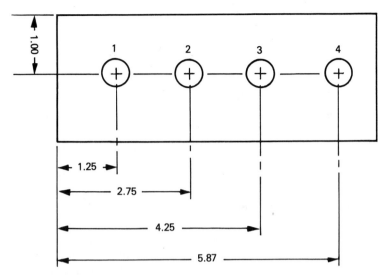

FIGURE 3-21
Absolute dimensioning.

sitional moves come from the same datum edge at all times, as opposed to an incremental system, where each succeeding move is an incremental distance from the last.

One advantage of absolute systems over incremental systems concerns positioning errors. If a positioning error occurs in an incremental system, all subsequent positions are affected, and all remaining moves are incorrect. When a positioning error occurs in an absolute N/C system, a particular location is in error but subsequent positions are not affected. This is because all dimensions and the corresponding positional moves are always based from the same zero or reference point. This is not to say, however, that the absolute system is superior to the incremental system. Such an attitude originated in the early days of numerical control because purchasers were forced to choose between an absolute system and an incremental system. Consequently, all detailed workpiece drawings had to be dimensioned to conform with the particular control mode, or the programmer was forced to make the translation when preparing the program tape and manuscript.

Both absolute and incremental systems have their logical areas of application, and neither is always right or wrong. There are certain applications in which both systems can be used most efficiently, sometimes even within the same program. Most controls today are capable of switching back and forth and working in either mode with just a simple instructional code inserted to make the change. With the adaptability of modern controls, the controversy over which is better is of little importance. In many cases, the burden of decision is placed directly on the programmer, who must have a thorough understanding of both modes and be able to make the best use of each.

FULL FLOATING ZERO

Some type of *zero shift* capability exists on most N/C machines. This capability implies shifting the zero location of the workpiece to any reasonable location on the machine table. Machines and controls range from older fixed-zero machines to full range offset (offsetting the zero location to any point on the machine table) and full floating zero. Modern machine tools today use full floating zero for shifting the machine zero.

With fixed-zero machines, there is no means of changing the zero location. The zero location is permanent and cannot be adjusted. Zero offset is used on fixed-zero machines, and the control *remembers* the permanent location of zero.

A *full floating zero* machine has no fixed reference point (zero point). For such a machine, the zero is established for each setup. When using an N/C system equipped with full floating zero, the operator may locate the workpiece at any convenient location on the machine table. Once the workpiece is set up or positioned on the table, the operator then obtains the alignment positions for the particular workpiece program from the program manuscript. In figure 3-22, the alignment positions in X and Y are X = 0 and Y = 0, or (0,0). These values are then dialed into the control. The operator depresses the cycle start button, and the machine rapid traverses to the location of the previous workpiece setup. The operator then manually moves the zero location of the machine table to

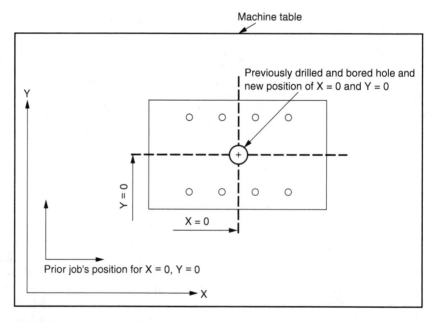

FIGURE 3-22
Full floating zero N/C system.

match the new setup's zero location without changing the actual readout on the control for X and Y. Aligning the machine tool merely operates the drive motors in X and Y, for example, but their signal does not enter the memory section of the control unit.

Once the new workpiece is "zeroed," or "trammed" in, the machine can be locked and synchronized with the control. The operator can then consistently run parts according to N/C or CNC commands relative to the manuscript alignment position and the convenient zero location.

Full floating zero greatly enhances actual machine spindle cutting time by reducing setup time. The programmer and operator gain flexibility because the programmer now can make "zero" any place on the machine table, enabling positive and negative programming.

REVIEW QUESTIONS

1. Compare a machinist's and programmer's method of producing a workpiece.

2. How is the Z-axis on an N/C machine determined?

3. Discuss the basic process that occurs within the control unit between data input and actual machine movement.

4. What are the primary differences between point-to-point and contouring N/C systems?

5. What is the difference between a closed-loop feedback system and an open-loop system?

6. Name the four types of drive motors used on N/C and CNC equipment.

7. Explain why the distinction between positioning and contouring N/C systems has become less significant.

8. What is the importance of the Cartesian coordinate system to numerical control systems?

9. What three axes are normally considered in N/C and CNC machining?

10. When programming for an N/C machine, is it necessary to write plus signs with all positive values? Why?

11. What is considered a positive Z movement? What is considered a negative Z movement?

12. What is a disadvantage of an incremental system that is an advantage of an absolute system?

13. What should a programmer know about absolute and incremental N/C systems?

CHAPTER 4

N/C Data Communication

OBJECTIVES

After studying this chapter, you will be able to:

- Identify the five different methods of N/C data input to the MCU
- Describe differences between direct and distributed numerical control
- Discuss EIA and ASCII coding systems
- List the five primary causes of N/C and CNC downtime

N/C DATA INPUT AND STORAGE

A variety of control systems have been developed since the beginning of numerical control. Consequently, a wide variety of different methods for N/C data *input* have also been developed. In order to standardize this N/C information, an EIA subcommittee was formed many years ago to recommend a set of standards that would be acceptable to control system manufacturers and machine tool manufacturers and users.

Several types of *input media* were tried as numerical control evolved. Initially, the most common types were punched cards, magnetic tape, and punched tape. Punched cards and magnetic tape (on open reels) proved highly impractical for shop use. Eventually, industry adopted a standard coded tape format in the form of punched tape because it was more suitable for shop environments.

Today, N/C data may be passed to the machine tool controller through five different means:

- Punched tape
- Cassette tape
- Floppy disk
- Manual data input
- DNC

Punched Tape

Tape readers (see figures 4-1 and 4-2) utilize punched tape input and are classified as mechanical or photoelectric (light) readers. They may read a single row of information at a time, or they may read a complete *block* of instructions.

Reader speeds will vary considerably; older mechanical readers were capable of reading approximately 60 characters per second, and photoelectric readers can read approximately 300 to 500 characters per second.

Photoelectric readers are the most commonly used because of their speed. Photoelectric readers operate by means of light beams that pass through the holes in the tape and impinge on a photocell. The light beams are then converted to electrical impulses and are passed on to the controller, providing smooth and continuous motion of the machine tool.

Punched tape is available in several different types of materials and colors. The tape materials are primarily grouped under three main headings: paper, Mylar (Du Pont's trade name for a tough plastic), and foil. There are, however, other combinations and variations of these three.

FIGURE 4-1
A typical CNC cabinet with a tape reader and tape reels. (Courtesy of Cincinnati Milacron Inc.)

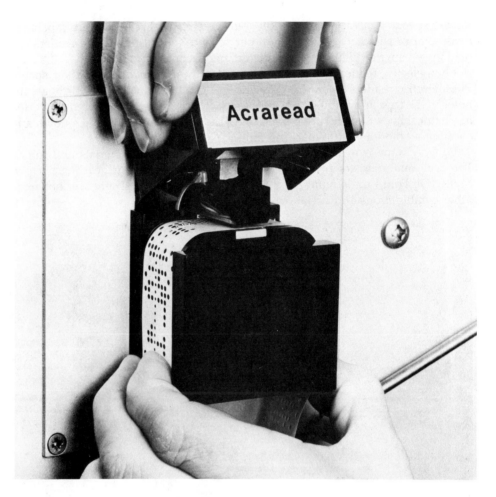

FIGURE 4-2
Loading a tape reader. (Courtesy of Cincinnati Milacron Inc.)

The specifications for the size of punched tape have varied tremendously since the early years of N/C development. Standardization became necessary in order to cut costs and establish common methods of programming between machine tool manufacturers. This was done through the cooperation of the EIA and the Aerospace Industries Association (AIA). Standardization covers two important categories: character coding and physical dimensions.

As shown in figure 4-3, the physical dimensions of the tape have been standardized—1 in. wide with eight tracks. It was determined that these tracks, or *channels*, were to run the length of the tape. The actual dimensions for thickness, hole spacing and size, and tolerances were also established at that time. Figure 4-4 shows all character codes and the appropriate punches for 1-in.-wide, eight-track tape (EIA-RS-244).

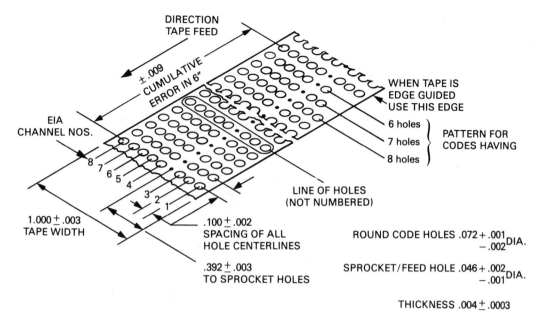

DIRECTION
TAPE FEED

±.009
CUMULATIVE
ERROR IN 6"

WHEN TAPE IS
EDGE GUIDED
USE THIS EDGE

EIA
CHANNEL NOS.

6 holes
7 holes
8 holes

PATTERN FOR
CODES HAVING

8 7 6 5 4 3 2 1

LINE OF HOLES
(NOT NUMBERED)

1.000 ± .003
TAPE WIDTH

.100 ± .002
SPACING OF ALL
HOLE CENTERLINES

ROUND CODE HOLES .072 + .001 DIA.
                              − .002

.392 ± .003
TO SPROCKET HOLES

SPROCKET/FEED HOLE .046 + .002 DIA.
                                − .001

THICKNESS .004 ± .0003

FIGURE 4-3
Standard 1-in.-wide, eight-track tape with dimensions and tolerances shown.

In the early days of N/C, actual tape preparation (punching) was done by manual tape preparation machines, similar to typewriters, called "flexowriters." Although N/C tape punching is still used by some manufacturers today, it is done by means of computer tape-punching equipment, as shown in figure 4-5. Today the medium has evolved into some type of electronic data storage such as disk or solid-state memory.

Cassette Tapes and Floppy Disks

Cassette tapes, similar to those found in automobiles and stereo equipment (see figure 4-6), are used today by some manufacturers for inputting and storing CNC part program data. Additionally, floppy disks, similar to those used on personal computers (see figure 4-7), are also a preferred choice of input and storage media for CNCs. Normal storage capacity ranges from 100 to 200 ft. of part program storage, but some double-density disks, for example, can hold considerably more. However, two common problems still exist for cassette tapes, floppy disks, and punched tape: storage and revision control. Regardless of the type of external storage media used, trying to store, manage, and keep up-to-date with the latest revision of part program changes is an administrative nightmare. The use of external storage media can also lead to the additional problems of possibly scrapping parts; causing tool, machine, or part damage; and potentially causing operator injury because of an outdated version of an N/C program being run.

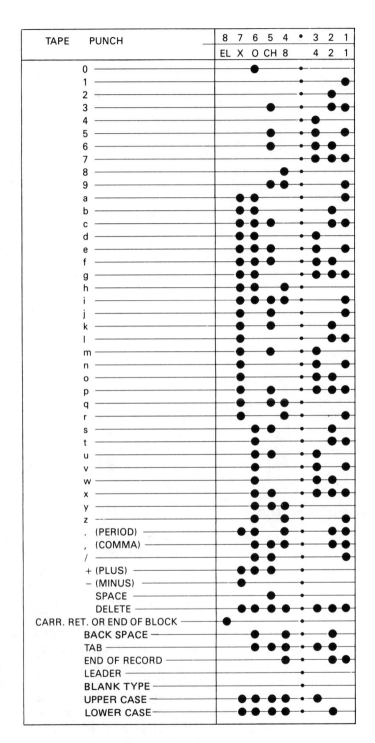

FIGURE 4-4
Character codes and punches for the EIA/BCD system (RS-244).

FIGURE 4-5
Computer tape-punching equipment. (Courtesy of Numeridex)

FIGURE 4-6
A typical cassette used for loading and storing CNC data.

FIGURE 4-7
A floppy disk used for
loading and storing CNC
data.

Manual Data Input

Manual data input (MDI) is a means of programming the machine tool through
the CNC keyboard and push buttons. MDI input, as shown in figure 4-8, can
range from just positioning the X-, Y-, and Z-axes to inputting a complete part
program and running that program from memory. MDI is completely operator-

FIGURE 4-8
Manual data input (MDI). (Courtesy of Okuma Machinery Inc.)

controlled and therefore subject to human error. Speed and accuracy depend entirely on the operator's ability to find, interpret, and input the correct information. All CNC machines today are capable of accepting and storing manual part programs.

CNC microprocessors have given birth to broad new programming aids in CNC that make the operator or programmer's job easier and quicker and reduce the length of (or eliminate completely) the part program. Generally, these programming aids consist of (1) subroutines permanently stored in CNC memory and (2) a shop floor programming language stored in CNC memory. Some CNC executive programs are written so that they will execute common subroutines such as bolt-hole circles and pocketing sequences from a single descriptive statement. Such capabilities are a definite aid when manual programming is being used. Both aspects of these programming aids will be discussed in greater depth later in the text.

DNC

Many CNCs in use today are wired directly to a central or personal computer for direct downloading of N/C data to the MCU. N/C programs can be generated, stored, and edited on the computer and then downloaded to each machine as needed. Such systems are called DNC for direct or distributed numerical control. A comparison of direct and distributed numerical control is illustrated in figure 4-9.

DNC systems bypass any punched tape, cassette, or floppy disk reader and use a standard RS-232-C communications port and coaxial cable for connecting the computer to the MCU. Older N/C and CNC controls do not have the RS-232-C communications port; they require a direct connection in place of the tape reader called a BTRI (behind-the-tape-reader interface). Direct numerical control systems download only a small portion of the program each time a part is run, feeding the machine with N/C data much like a tape reader. Distributed N/C systems download the entire program into the MCU's memory.

Typically, the CNC has an internal memory for program storage. So, depending on the MCU and its input media, it is possible to "read in" and store part program data from punched tape, cassette tape, floppy disk, MDI, or distributed N/C and run the program continuously from memory. An example is illustrated in figure 4-10. Storing 200 ft. of punched tape is common for many CNCs, but some modern controls will take up to 2,000 ft. If the CNC has a 20-MB hard disk, for example, the MCU can hold 75,000 ft. or more of taped program storage.

Primary advantages of distributed N/C systems include the following:

1. Punched tape, floppy disks, and cassette tapes are eliminated as media for storing and loading programs.

2. Program revisions and tool lists are more easily managed.

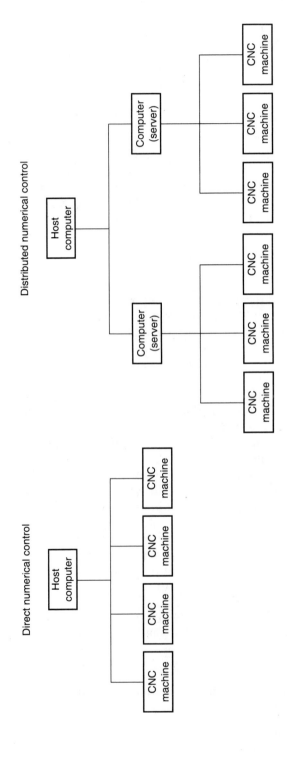

FIGURE 4-9

A comparison of direct and distributed numerical control.

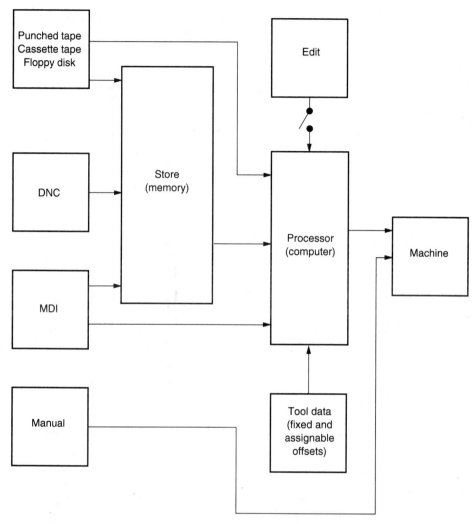

FIGURE 4-10
How a CNC unit functions in conjunction with its microcomputer.

3. Part programs have a centralized and on-line location for access and retrieval.

4. Program security is improved.

5. Optimized "on-machine" changes to the part program can be "uploaded" to the central computer and stored after the part is run.

6. The DNC system can be used for other plant-wide data collection and distribution.

BCD (RS-244)									CHARACTER	ASCII (RS-358)								
		●		•					0			●	●	•				
				•				●	1	●		●	●	•				●
				•			●		2	●		●	●	•			●	
			●	•			●	●	3			●	●	•			●	●
				•		●			4	●		●	●	•		●		
			●	•		●		●	5			●	●	•		●		●
			●	•		●	●		6			●	●	•		●	●	
				•		●	●	●	7	●		●	●	•		●	●	●
				•	●				8	●		●	●	•	●			
			●	•	●			●	9			●	●	•	●			●
	●	●		•				●	a	●	●	●		•				●
	●	●		•			●		b	●	●	●		•			●	
	●	●	●	•			●	●	c		●	●		•			●	●
	●	●		•		●			d	●	●	●		•		●		
	●	●	●	•		●		●	e		●	●		•		●		●
	●	●	●	•		●	●		f		●	●		•		●	●	
	●	●		•		●	●	●	g	●	●	●		•		●	●	●
	●	●		•	●				h	●	●	●		•	●			
	●	●	●	•	●			●	i		●	●		•	●			●
	●		●	•				●	j		●	●		•	●		●	
	●		●	•			●		k	●	●	●		•	●		●	●
	●			•			●	●	l		●	●		•	●	●		
	●		●	•		●			m	●	●	●		•	●	●		●
	●			•		●		●	n	●	●	●		•	●	●	●	
	●			•		●	●		o		●	●		•	●	●	●	●
	●		●	•		●	●	●	p	●	●	●	●	•				
	●		●	•	●				q		●	●	●	•				●
	●			•	●			●	r		●	●	●	•			●	
		●	●	•			●		s	●	●	●	●	•			●	●
		●		•			●	●	t		●	●	●	•		●		
		●	●	•		●			u	●	●	●	●	•		●		●
		●		•		●		●	v	●	●	●	●	•		●	●	
		●		•		●	●		w		●	●	●	•		●	●	●
		●	●	•		●	●	●	x		●	●	●	•	●			
		●	●	•	●				y	●	●	●	●	•	●			●
		●		•	●			●	z	●	●	●	●	•	●		●	
	●	●	●	•	●	●			.			●		•	●	●	●	
		●	●	•	●	●	●		,	●		●		•	●	●		
		●	●	•				●	/	●		●		•	●	●	●	●
	●	●	●	•					+			●		•	●		●	●
	●			•					−			●		•	●	●		●
			●	•					Space	●		●		•				
●	●	●	●	•	●	●	●	●	Delete	●	●	●	●	•	●	●	●	●
●				•					Car. Ret. or End of Block	●				•	●	●		●
		●		•	●	●			Back Space	Not Assigned								
	●	●	●	•	●	●			Tab					•	●			●
		●	●	•	●	●		●	End of Record	Not Assigned								
				•					Leader	Not Assigned								
●	●	●	●	•	●				Upper Case	Not Assigned								
●	●	●	●	•				●	Lower Case	Not Assigned								

BCD - Binary Coded Decimal system
ISO - International Standards Organization
ASCII - American Standard Code for Information Interchange
The systems ISO and ASCII are the opposite of BCD in that they are even-parity coded.

FIGURE 4-11
A comparison of EIA/BCD and ASCII coding systems.

EIA AND ASCII CODE

The Electronics Industries Association developed the *binary-coded decimal* (BCD) system for coding tapes several years ago. In this binary system of number representation, each decimal digit is represented by a group of binary digits that correspond to a character. It incorporates the best features of both the binary and decimal systems and gives a compact and simple method of converting decimal dimensions into the analog voltage ratios that control the machine tool.

The *American Standard Code for Information Interchange* (ASCII), another coding system that was developed years ago, is now the standard coding system used throughout the computer and communications industries. It was compiled by a committee made of several different groups working with the United States of America Standards Institute. This group, now known as the American National Standards Institute (ANSI), had the overall objective of obtaining one coding system to serve as an international standard for all information-processing and communications systems.

There are several coding differences between EIA/BCD and ASCII. However, modern CNCs can automatically detect and adjust for either EIA or ASCII input. ASCII provides coding for both uppercase and lowercase letters, whereas BCD codes are the same for both. The 10 digit codes (0 through 9) in ASCII are the same as in BCD coding, but with ASCII holes are punched in two additional tracks to identify numbers and certain symbols. The ASCII letter codes, however, are quite different from those used in BCD. A comparison of the two systems is shown in Figure 4-11.

TYPES OF TAPE FORMATS

Tape format refers to the general sequence and arrangement of coded information on a punched tape. This information conforms to EIA standards and appears as *words* based on individual codes and written in horizontal lines, as shown in figure 4-12. For example, there are five *words* that make up a *block* (one instruction) for this particular tape format. The most common type of tape format in current use is decimal point programming. However, an earlier format still in use today is the word address format (also known as the interchangeable or compatible format). Fixed sequential, tab ignore, and tab sequential are older formats that are obsolete today. Examples of these formats are illustrated in figure 4-13.

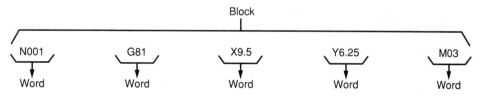

FIGURE 4-12
A block of information and the individual words.

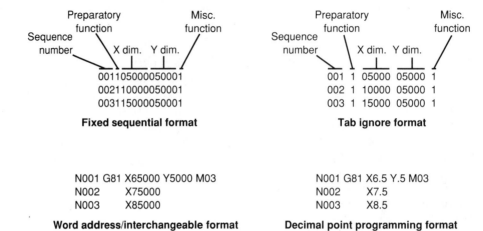

FIGURE 4-13
Examples of tape formats.

The word address format, standardized by the EIA years ago, uses a letter address to identify each separate word. By assigning an alphabetical code to each coordinate and function word, the block format becomes more flexible. Letter addresses are used for word and register identification. They minimize the amount of N/C data storage because only codes that change from one block to the next need to be programmed; repeated codes can be omitted. In addition, words do not have to appear in a rigid order, and tapes are more interchangeable with machines in the same class. Word addresses typically conform to the ANSI/EIA RS-274-D standard shown in table 4-1. This standard is followed by most CNC equipment manufacturers.

TABLE 4-1   ANSI-EIA RS-274-D
Standard Word Addresses

A	Rotation about the X-axis
B	Rotation about the Y-axis
C	Rotation about the Z-axis
F	Feed rate commands
G	Preparatory functions
I	Circular interpolation X-axis offset
J	Circular interpolation Y-axis offset
K	Circular interpolation Z-axis offset
M	Miscellaneous commands
N	Sequence number
O	Sequence number for secondary axis commands
R	Arc radius
S	Spindle speed
T	Tool number
X	X-axis data
Y	Y-axis data
Z	Z-axis data

It should be pointed out that even though a general tape format is followed, the specific tape format for a particular machine tool and its specific control type must also be followed. For example, a CNC lathe and a CNC machining center are two different and unique N/C machines. Both can be programmed using the decimal point programming format, but the N/C lathe tape input will not work in the machining center, and vice versa. This is because each machine and control system has its own set of words that are recognizable. If the N/C control is fed unrecognizable information (for example, if an N/C lathe tape is accidentally loaded in a machining center control), a system failure will occur, and the machine will not make any movements. Therefore, it is important to understand the specific tape format for the particular machine and control type being programmed.

N/C AND CNC DOWNTIME

The primary causes of N/C, CNC, and other computer equipment breakdowns are

1. Heat
2. Power disturbances
3. Contaminated air
4. Vibration
5. Oxidation

The MCU and associated machine tool downtime they cause can in many cases easily be prevented if proper planning and prevention methods are followed. Case histories clearly indicate that controlling the environment within which N/Cs and CNCs operate is the most cost-effective way to increase uptime and reduce maintenance.

Heat

Electronic controllers and computer systems use a variety of integrated circuits (ICs), transistors, diodes, and other components—each with its own heat tolerance level. Each component operates in a specific low/high temperature range outside of which it fails. Therefore, MCU cabinets housing these components must maintain a temperature that stays well within the range for each component. In addition, temperature inside the cabinets should not fluctuate too much because fluctuations can cause servodrive misalignments as well as condensation and oxidation.

MCU cabinet air conditioners should maintain a specific temperature between 75 and 100°F. When shop air raises cabinet temperature above 100°F, the air conditioner should step up cooling. Temperature gages on MCU cabinets can provide maintenance personnel a clearly visible and accurate measure of conditions inside each cabinet.

Power Disturbances

Electrical power supply disturbances, usually in the form of transients and spikes, are a major source of electronic and computer equipment failure. These disturbances can be eliminated through currently available suppression devices. Suppressors lessen transient electrical energy once it exceeds the nominal peak voltage of the electronic equipment.

Contaminated Air

Many MCU cabinets are insufficiently sealed against shop air. Layers of oily dirt or black iron filings can accumulate on circuit boards and cause permanent hardware damage. The dust and grit enter cabinets through wire ducts and vent panels or around conduits, switches, and door edges if the cabinet is improperly sealed.

Shop air circulated through cabinets is a major cause of dirt accumulation inside cabinets. To keep dirt out of the cabinet, all ventilation systems must be sealed off; the air conditioner is required to both cool and recirculate air. Doors and edges should be well sealed off with commercially available foam-rubber strips, and cable holes should be sealed with nonflammable sealing material.

Vibration

In addition to dirt, physical vibration continues to be another major reason for controller and computer failure. Circuit boards can gradually vibrate out of their sockets, causing intermittent or lost connections and consequently equipment breakdowns. Vibration also shears components as well as lead and plug socket connections over time. Shock-absorbing pads under cabinets can help reduce vibration.

Oxidation

Electrical/electronic connections can also be rendered nonfunctional by normal aging or oxidation of a component. Solutions are available to clean oxidized connectors, but they may actually cause the connectors to corrode faster or form a thin film, thereby attracting dirt.

Summary

Computer and electronic equipment work best in the proper environment. In a business-computer installation, floors are often raised and insulated, and a lot of effort goes into planning the right air and power utilization levels. However, shop applications do not typically receive the same level of consideration, even though industrial systems use the same electronic components as business systems and are exposed to a much harsher environment.

DATA MANAGEMENT AND ADVANCED CONNECTIVITY

Modern data demands in manufacturing consist of not only part programs, but a wide range of information about activity on the factory floor. Such information as process routings, setup sheets, part geometry graphics, statistical process control data, tool and material descriptions, work order schedules, and other factory management data is beginning to flow through DNC systems. This change is being driven by at least two factors. First, there is an increasing need for manufacturing management to measure and increase manufacturing efficiency in order to remain competitive in the world marketplace. Second, since a DNC system is a communication system attached to a machine tool, it is the logical choice for a data collection and communication point linked to a manufacturing data repository. The alternative is to introduce additional systems and terminals that create data-entry redundancies; increase chances of input error; add additional system support, training, cost, and familiarity requirements to operating personnel; and can create multiple databases and data management burdens.

Although many CNCs are currently in use, the real requirement today is how to connect, in real time, all sources of information internally and sometimes externally. Essentially this involves optimizing for the whole, attempting to use as few databases as possible, and not suboptimizing for "islands of automation." Advanced connectivity provides management and operating personnel with the information needed to make quick, real-time decisions based on accurate information and to share that information with everyone for the benefit of all. Applications must be able to talk to each other in order to share accurate and current information and avoid redundant data-entry, storage, and management issues and problems. This requirement is a data integration issue and is the essence of computer integrated manufacturing (CIM.) CIM is discussed in more detail in Chapter 11.

REVIEW QUESTIONS

1. Name the five different types of N/C data input to the MCU, and briefly describe each.
2. Briefly compare and contrast direct versus distributed N/C.
3. List four advantages of DNC (distributed N/C).
4. What two groups were responsible for the progress made in tape standardization?
5. What is the standard coding system used today throughout the computer and communication industries?
6. Explain what is meant by tape format.
7. What is the most commonly used tape format today?
8. What is the purpose of the word address for N/C and CNC words?
9. What is a BTRI? When and why are they used?
10. What are the five primary causes of N/C and CNC downtime?

CHAPTER 5

Simple Part Programming: Conventions and Examples

OBJECTIVES

After studying this chapter, you will be able to:

- Describe basic part programming methodology
- Identify various N/C codes and functions
- Discuss programming-related machine tool movements resulting from N/C input
- Demonstrate a knowledge of various auxiliary function commands and their importance

FUNCTIONS CONTROLLED BY N/C

Generally, an N/C or CNC program manuscript, regardless of machine type, consists of the heading, machine tape information, and operator information. The heading contains such identification information as part name and number, drawing and fixture number, date, and programmer's name. Machine tape information contains all information necessary for proper machine operation, such as function and miscellaneous codes; X, Y, and Z positions; and feed rate and spindle speeds. This information will vary depending on the particular machine tool being used. The operator information section of an N/C program contains position number, depth of cut, operation description, and cutting tool information. This section also includes information on setup and machine alignment, as well as any other information the programmer feels the operator should know. This information is used strictly as an aid to the operator for a better understanding of how to prepare for and process the job.

The specific N/C words that control N/C are part of the machine tape information and are printed on a manuscript for operator and programmer reference. These words consist of the following: sequence numbers; preparatory and miscellaneous functions; X, Y, and Z coordinate information; spindle speeds; and feed rate. Each word consists of alphanumeric codes that relate to a specific register in the machine control unit; these codes either are informational only or cause an appropriate machine tool movement or action to occur. Each of the basic N/C words is discussed in more detail in this chapter.

Sequence Number

Normally, the first word in a block of information is the *sequence number*. This word, among others, appears on the operator's CNC cathode ray tube (CRT) screen. Its primary purpose is to identify each block of information so that it can be distinguished from the rest and to indicate the block of tape being performed. The sequence number is usually a three- or four-digit word. The sequence number is preceded by the letter H, O, or N. Sequence numbers are normally progressive and informational rather than functional. Examples would be O001, N002, and N003.

H or O blocks, normally used on older controls, were used for the first block of information in every program, after every tool change, and for alignment and realignment blocks. The letter N was was used for all other sequence blocks. On most CNCs today, all sequence numbers use the letter N to address all blocks.

H and O blocks were also frequently referred to during tape searches. The search feature, incorporated into most CNC units, allows the system to search the program, by mean of operator intervention, for a particular sequence number and to stop when the number is found. Most modern CNCs can search for any specific block in the part program. The search feature allows information to be temporarily bypassed or recalled as needed.

X and Y Words

Regardless of the type of N/C machine being programmed, coordinate information is necessary for the machine tool to position itself. This information may be expressed using X, Y, and Z words. However, in this preliminary discussion, we will concentrate on X and Y words.

Because decimal point programming is the most common format in use today, coordinate input normally uses only as many significant digits as required. The X and Y word addresses still need to be programmed, but insignificant digits may be dropped. For example, if an X dimension of 5.5 in. needs to be programmed, it would be entered as X5.5. It would not be necessary to program X5.500. Appending the two extra zeros would not make the programmed dimension any more accurate.

Many CNC machines still use the interchangeable, or compatible, format. X and Y words in this format are seven-digit numbers, along with the sign of the number, preceded by the letter X to indicate the X-axis and the letter Y to indicate the Y-axis. The X and Y words are written as X±1******* and Y±1*******. The position of the decimal point in the coordinate information is fixed in this tape format to allow four places to the right of the decimal point. The word *X+0043750* specifies a coordinate of 4.3750 along the X-axis and is accepted by the control system. Most controls have the ability to retain the sign of the programmed word. Because of this feature, and depending on the machine and control system, the sign of the number need be programmed only when it changes from the previous block of information. On older controls using an H

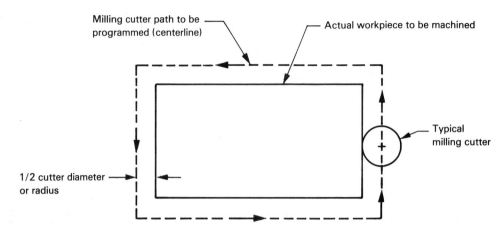

Milling cutter path to be programmed (centerline)

Actual workpiece to be machined

Typical milling cutter

1/2 cutter diameter or radius

FIGURE 5-1
Importance of accounting for the cutter radius when centerline locations are being programmed.

or O block, however, the sign of the number should always be programmed to ensure the correct sign input when starting a series of operations.

It is important to remember when one is programming X and Y words that the actual centerline of the cutter is always what needs to be programmed, as shown in figure 5-1. Establishing X and Y values for point-to-point (hole pattern) operations is relatively easy because the direct X and Y word input consists of specific locations, without continuous cutting between the locations. Milling, in contrast, is more involved because the cutter radius must always be taken into account when the X and Y (centerline) locations are being programmed.

Leading and Trailing Zero Suppression

Earlier controllers required a certain number of digits in each block. If a digit was not needed, the space was filled with a place-holding zero. Depending upon whether leading or trailing zeros were required, zeros were placed before or after significant numbers to fill the full block. Today, CNC controllers do not require leading and trailing zeros; decimal point programming is the currently accepted standard, and decimals are programmed into the words.

However, many N/C and CNC controllers still employ the interchangeable or compatible format with leading *zero suppression*. The coordinate information in the word-addressed X and Y registers enters, in most cases, in a right-to-left sequence, as shown in figure 5-2. The decimal point is automatically positioned four places to the right of the X or Y word.

Study figure 5-2. Although there are seven digit positions available, only five are needed. The word could be written as X+0041250, and the two preceding zeroes could be programmed and entered into the control. However, because the words enter the registers from right to left, the two *leading zeroes*, to the left of the *significant digits*, are insignificant and do not need to be programmed. (The control assumes the same sign for successive words unless the sign is

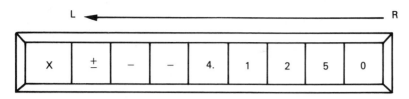

FIGURE 5-2
Coordinate information with fixed decimal location and right-to-left
registration order.

explicitly changed in the word.) Therefore, they can be suppressed (that is, omitted). Either format would be acceptable to the control and machine, but the leading zeroes are insignificant and have no effect on the programmed word. This omission of insignificant digits is called *leading zero suppression*.

Preparatory Functions

The *preparatory function*, or cycle code, is a two-digit number (00–99) preceded by the word address letter G. This code, referred to as the *G code*, determines the mode of operation of the system. It will cause different actions or operations to occur on the machine. There have been attempts to standardize G codes among CNC manufacturers, and many codes are common, but nothing has been officially finalized.

There are four main categories of G codes. The respective categories are used to do the following:

1. Select a movement system
2. Select a measurement system (metric or English)
3. Program compensation for variation in tool diameters and lengths
4. Select preset sequences of events known as canned cycles

Linear moves involve straight-line movements of a machine tool. The production of straight line by the simultaneous movement of two or more axes is called *linear interpolation* (interpolation refers to finding a value between two given values). G00 and G01 are the two primary commands used for linear moves on most CNC equipment (see figure 5-3). G00 is used for point-to-point positioning at the maximum rapid traverse feed rate of the machine tool. G01 is used to position two or more axes simultaneously at a programmed feed rate.

A *canned cycle* is a combination of machine moves resulting in a particular machining function such as drilling, milling, boring, and tapping. One cycle code number in a program may cause as many as six, seven, or more distinct machine movements to occur. Without canned cycles, these movements would normally require several blocks of programming. Most control manufacturers today have both canned and noncanned cycles as part of their standard control packages.

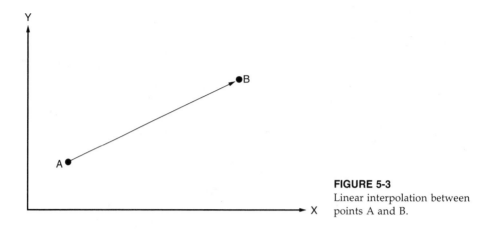

FIGURE 5-3
Linear interpolation between points A and B.

Consider a canned cycle that consists of the following sequence of six operations:

1. Positioning along the X- and Y-axes
2. Rapid traverse to point R
3. Drilling
4. Operation at bottom of hole (if any)
5. Retraction to point R
6. Rapid traverse to initial point

Figure 5-4 illustrates a schematic of this sequence. An example of code that could be used to program this sequence is as follows:

```
N001   G55   G90 G00 X0 Y0
N002   G43   H2 Z.5
N003   M03   S1000
N004   G81   R.05 Z−.1875 F20.
```

In the second line of code, Z.5 sets up the initial point at a Z height of 0.5 in. (that is, 0.5 in. is the first Z position called). In the fourth line of code, R.05 says that point R is 0.050 in. above the Z-axis zero. In general, the R word establishes a predetermined "rapid to" point just above the work surface. The value after R is changeable on some controls and fixed at 0.100 in. on others. The R word will be discussed in more detail later in the text.

The codes G98 and G99 are used to specify the retraction point of the cutting tool after the desired Z depth is reached (see figure 5-5). If neither G98 nor G99 is specified on the canned cycle line of information, initial-point return is performed. If return to point R is desired, G99 must be programmed on the canned cycle line of information.

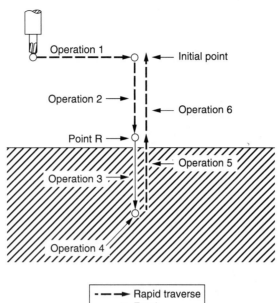

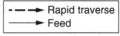

FIGURE 5-4
A canned cycle consisting of a
sequence of six operations.

---▶	Rapid traverse
──▶	Feed

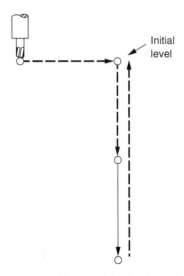

Return to inital level (G98)

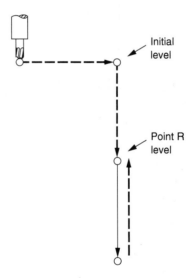

Return to point R level (G99)

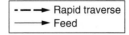

---▶	Rapid traverse
──▶	Feed

FIGURE 5-5
The codes G98 and G99 specify the retraction point of the cutting tool after the desired
Z depth is reached.

EXAMPLES OF CANNED CYCLES

This section will explain individual canned cycles as well as the individual variables on each line of information. Program examples will be given in the proper line format for each canned cycle.

G73: High-speed Peck-drilling Cycle

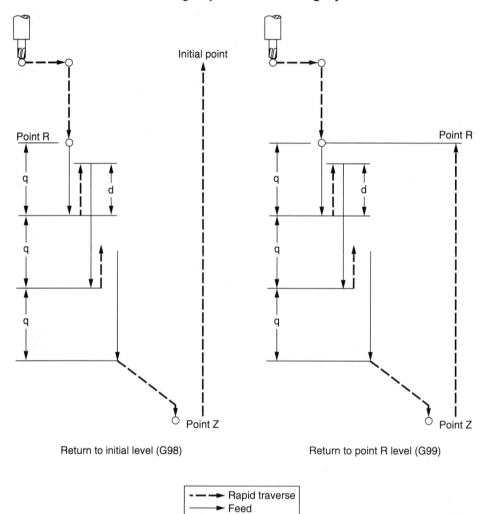

Return to initial level (G98)   Return to point R level (G99)

- - - ▶ Rapid traverse
——— ▶ Feed

The code G73 is used when holes are being drilled to a specific depth. In this drill cycle, a Q word is present. The Q word value represents the incremental cut-in amount that occurs before the drill rapid retracts. After each incremental cut, the drill will rapid retract a certain amount (indicated by the letter *d* in the G73 peck-drilling cycle). This retract amount is not programmed in the canned cycle. The value is stored in the control parameters and is automatically called up when G73 appears in a program being run. The retract amount is usually set at approximately 0.030 in., which is enough to break off a drill chip.

FORMAT EXAMPLE

G73 G99 X.5 Y2.5 Q.25 R.05 Z−1.5 F20.

G73	= high-speed peck-drilling cycle
G99	= rapid retract to point R
X.5 Y2.5	= coordinates along X- and Y-axes
Q.25	= 0.25-in. depth for each incremental cut
R.05	= point R specification of 0.05 in. above part surface (Z zero)
Z−1.5	= drilling depth of 1.5 in. in −Z direction
F20	= feed rate of 20 IPM

G74: Left-hand Tapping Cycle

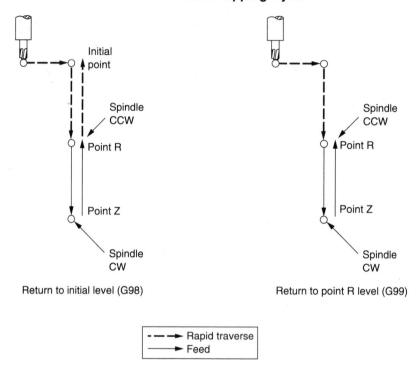

Return to initial level (G98)   Return to point R level (G99)

- - - ►	Rapid traverse
——►	Feed

The G74 canned cycle automatically turns the spindle on in the counter-clockwise direction to accommodate left-hand tapping (see the above figure). Upon reaching its programmed Z depth, the spindle reverses and retracts at a programmed feed rate.

FORMAT EXAMPLE

G74 G99 X1. Y.5 R.05 F5. Z−1.5

G74	= left-hand tapping cycle
G99	= rapid retract to point R
X1. Y.5	= coordinates along the X- and Y-axes
R.05	= point R specification of .05 in. above part surface (Z zero)
F5.	= feed rate of 5 IPM
Z−1.5	= feed 1.5 in. deep

Note: When the G74 canned cycle code is being used, the feed rate and spindle speed must be of corresponding values in order for this cycle to be performed properly. During the tapping canned cycle, any change by the feed rate override (the ability of the operator to override a programmed feed rate at the CNC control) will be ignored. When the G74 cycle is in process, the operation will not stop until the cycle is complete, even if the feed hold is applied.

G76: Fine Boring Cycle

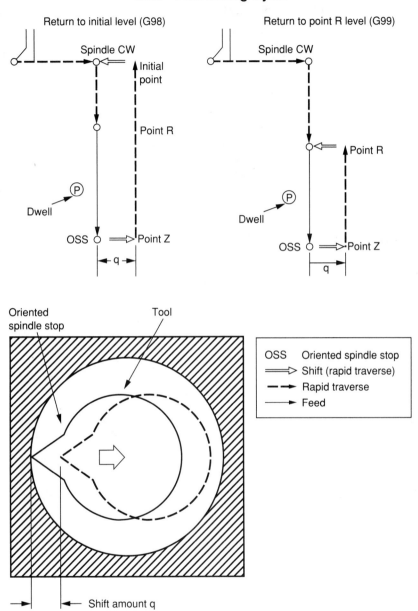

(See following page)

The G76 canned cycle is used when a hole being bored requires a fine sidewall surface finish. The spindle stops at the bottom of the hole, and the boring bar point is oriented to move away from the cutting surface (see the G76 figure). The programmed Q value indicates that stepover amount. Moving the cutting tool away from the hole sidewall prevents scratching as the tool retracts back to the starting position.

FORMAT EXAMPLE

G76 G99 X2. Y1. R.05 Q.125 F10. Z−1.5

G76	= fine-boring cycle
G99	= rapid retract to point R
X2. Y1.	= coordinates along the X- and Y-axes
R.05	= point R specification of 0.05 in. above part surface (Z zero)
Q.125	= stepover of 0.125 in. away from sidewall before retracting
F10.	= feed rate of 10.0 IPM
Z−1.5	= feed 1.5 in. deep

Note: The shift direction at the bottom of the hole may be set for any +X, −X, +Y, and −Y direction. The direction of offset is generally set in the control parameters.

G81:   Center-drilling, or Spot-drilling, Cycle

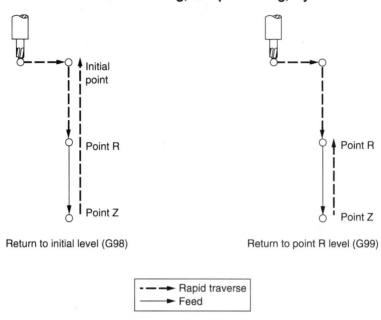

Return to initial level (G98)   Return to point R level (G99)

- - -► Rapid traverse
———► Feed

The G81 canned cycle is generally used for spot drilling, or center drilling, a number of holes at various locations. The center drill will rapid traverse to point R or the initial point, feed to the programmed depth, and rapid retract to point R or the initial point (see the above figure).

FORMAT EXAMPLE

G81 G99 X2.5 Y0 R.05 Z−.1875 F15.

G81	= center-drilling or spot-drilling cycle
G99	= rapid retract to the point R
X2.5 Y0	= coordinates along the X- and Y-axes
R.05	= point R specification of 0.05 in. above part surface (Z zero)
Z−.1875	= drilling depth of 0.1875 in. in −Z direction
F15.	= feed rate of 15 IPM

G82: Drilling and Counterboring Cycle

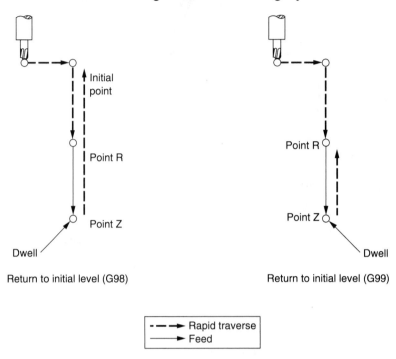

Return to initial level (G98) Return to initial level (G99)

- - - ▸	Rapid traverse
——▸	Feed

The G82 drilling and counterboring cycle operates basically the same as the G81 cycle except that a dwell is performed when the programmed Z depth is reached (see the above figure). The dwell ensures that the drill or boring bar has reached the final Z depth. The dwell allows the drill or boring bar to rotate a full 360° at the bottom of the hole. This rotation produces a finished and more accurate Z depth dimension.

FORMAT EXAMPLE

G82 G99 X2. Y0 R.05 P5. Z−.75 F20.

G82	= drilling, counterboring cycle
G99	= rapid retract to point R
X2.Y0	= coordinates along the X- and Y-axes
R.05	= point *R* specification of 0.05 in. above part surface (Z zero)
P5.	= dwell time in seconds
Z−.75	= drilling depth of 0.75 in. in −Z direction
F20.	= feed rate of 20 IPM

Note: Dwell is specified in seconds (from 0.01 to 99999.999).

G83: Peck-drilling Cycle

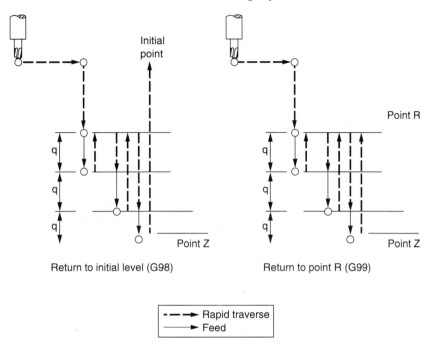

Return to initial level (G98)

Return to point R (G99)

Initial point

Point R

Point Z

Point Z

- - - ▶ Rapid traverse
———▶ Feed

The G83 canned cycle drills holes to a specified Z depth. In this canned cycle the Q word is also present. The value that follows Q is the incremental cut-in value. The drill will cut in this amount until it reaches the programmed Z depth (see the above figure). This canned cycle differs from the G73 canned cycle in that each time the drill completes an incremental cut, it retracts to the programmed R point.

FORMAT EXAMPLE

G83 G99 X3. Y2. Z−2.5 Q.15 R.05 F20.

G83	= peck-drilling cycle
G99	= rapid retract to point R
X3. Y2.	= coordinates along the X- and Y-axes
Z−2.5	= drilling depth of 2.5 in. in −Z direction
Q.15	= incremental drilling depth of 0.15 in.
R.05	= point R specification of 0.05 in. above part surface (Z zero)
F20.	= feed rate of 20 IPM

Note: The G83 peck-drilling cycle is primarily used when deep holes are being drilled. The retraction of the drill back to the R point helps relieve the drill of any chips that may become lodged in the drill flutes during operation.

G84: Right-hand Tapping Cycle

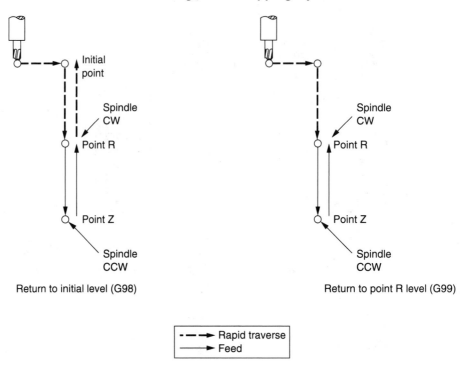

Return to initial level (G98)

Return to point R level (G99)

Rapid traverse
Feed

The G84 canned cycle is used when right-hand threads are being tapped. The spindle rotation changes from a clockwise to a counterclockwise direction after reaching the programmed Z depth (see the above figure).

FORMAT EXAMPLE

G84 G99 X0 Y0 R.15 Z−.625 F5.

G84	= right-hand tapping cycle
G99	= rapid retract to point R
X0 Y0	= coordinates along the X- and Y-axes
R.15	= point R specification of 0.15 in. above part surface (Z zero)
Z−.625	= programmed depth of .625 in. in −Z direction
F5.	= feed rate of 5 IPM

Note: When the G84 canned cycle is being used, the feed rate and spindle speed must be of corresponding values in order for this cycle to be performed properly. During the tapping canned cycle, any change by the feed rate override will be ignored. When the G84 cycle is in process, the operation will not stop until the cycle is complete, even if the feed hold is applied.

G85: Boring Cycle

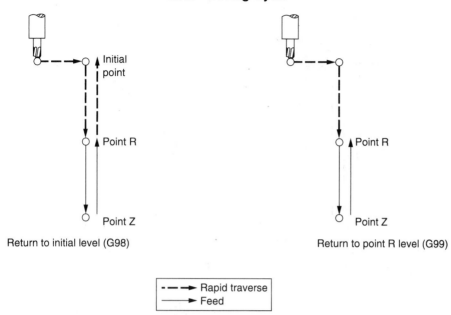

Return to initial level (G98)   Return to point R level (G99)

- - → Rapid traverse
——→ Feed

The G85 canned cycle is a boring cycle in which the cutter feeds to a programmed Z depth and back to a programmed R point (see the above figure).

FORMAT EXAMPLE

G85 G99 X4. Y2.5 R.05 Z−2.0 F20.

G85	= boring cycle
G99	= rapid retract to point R
X4. Y2.5	= coordinates along the X- and Y-axes
R.05	= point R specification of 0.05 in. above part surface (Z zero)
Z−2.0	= programmed depth of 2.0 in. in −Z direction
F20.	= feed rate of 20 IPM

G86: Boring Cycle

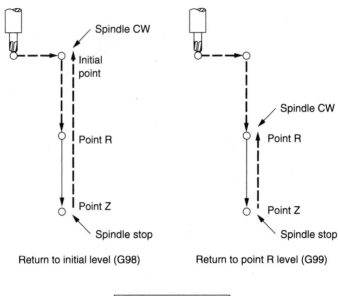

Return to initial level (G98) Return to point R level (G99)

- - - → Rapid traverse
———→ Feed

In the G86 boring canned cycle, the tool feeds to the programmed Z depth, the spindle turns off, and the tool retracts to the R point (see the above figure).

FORMAT EXAMPLE

G86 G99 X2.0 Y0 R.2 Z−4. F10.

G86	= boring cycle
G99	= rapid retract to point R
X2.0 Y0	= coordinates along the X- and Y-axes
R.2	= point R specification of .2 in. above part surface (Z zero)
Z−4.	= programmed depth of 4 in. in −Z direction
F10.	= feed rate of 10 IPM

G88: Boring Cycle

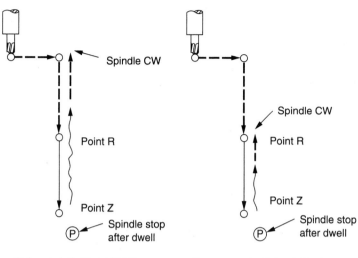

Return to initial level (G98)   Return to point R level (G99)

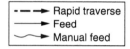

In the G88 canned cycle, the cutting tool rapid traverses to the R point and feeds down to the programmed Z depth (see above figure). After reaching the Z depth, the spindle stops and the system enters a halt state. During the halt state, the cutting tool can be manually removed from the hole by switching to manual mode. To ensure safe operation, make sure to manually retract the tool before continuing operation. When restarting in the memory mode, the tool will return to either the initial point (for G98) or the R point level (for G99).

FORMAT EXAMPLE

G88 G99 X3. Y2.2 R.075 Z−1.22 F10.

G88	= boring cycle
G99	= rapid retract to point R
X3. Y2.2	= coordinates along the X- and Y-axes
R.075	= point R specification of 0.075 in. above part surface (Z zero)
Z−1.22	= programmed depth of 1.22 in. in −Z direction
F10.	= feed rate of 10 IPM

G89: Boring Cycle

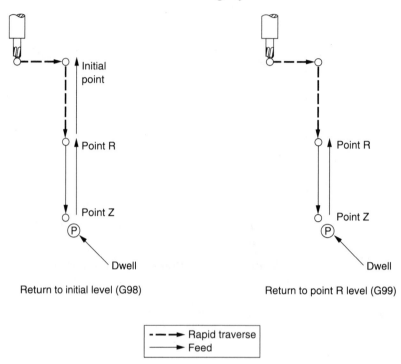

Return to initial level (G98)

Return to point R level (G99)

– – →	Rapid traverse
—— →	Feed

In the G89 canned cycle, the tool rapid traverses to the R point and then feeds down to the programmed Z depth (see the above figure). Once it reaches the programmed Z depth, a dwell is performed, and the tool feeds back to either the R or initial point.

FORMAT EXAMPLE

G89 G99 X1.75 Y0 R.1 Z−2.5 P5. F20.

G89	= boring cycle
G99	= rapid retract to point R
X1.75 Y0	= coordinates along the X- and Y-axes
R.1	= point R specification of 1 in. above part surface (Z zero)
Z−2.5	= programmed depth of 2.5 in. in −Z direction
P5.	= dwell of 5 sec.
F20.	= feed rate of 20 IPM

G80: Cancel Drill Cycle

The G80 code is used to cancel all programmed canned cycles. This code is generally called at the beginning of a program and after completion of the canned cycle operation.

FORMAT EXAMPLE

G55 G90 G40 G0 X0 Y0

The following code calls for a canned cycle to be performed at three locations before the drill cycle is canceled:

```
G83 G99 X2.5 Y0 R.05 Z−.5 F20. M03 S500
X2.0 Y1.
X1.5 Y1.2
G80
```

The tool changes and the program continues after the G80 line.

* * * * * *

The following are important rules concerning canned cycles:

1. The spindle must be rotating (via codes M03 or M04) before a canned cycle is called.

2. A canned cycle may be canceled by calling codes G00 to G03 (these codes will override the programmed canned cycle).

3. Canned cycle variables may be changed on consecutive blocks of information; the canned cycle call code need not be repeated. Changed variables will stay in effect for consecutive blocks of information.

4. Tool offset commands G45 through G48 (discussed later in the text) are ignored in the canned cycle mode.

5. Do not press the ORIGIN button on the control during canned cycle operation. (It will change the machine origin position and could cause an accident.)

6. When tool-length offset (codes G43, G44, G49) is specified during the canned cycle mode, the offset applies when the tool is positioned at point R.

Feed Rates and Spindle Speeds

Feed rates govern the amount and rate of metal removal for a particular tool and type of workpiece material to be machined. Feed rates are normally measured and programmed in inches per minute (IPM) but can also be programmed in inches per revolution (IPR), millimeters per minute (mm/min.), or millimeters per revolution (mm/rev.).

For most machine tool and control manufacturers, the feed rate (F) word is directly programmed using decimal point programming. The feed rate code is an F word address followed by the feed amount. A feed rate of 4.5 IPM, for example, would be programmed as F4.5. Machining centers are usually programmed in inches per minute. Turning centers are normally programmed in inches per revolution. A feed rate of .007 IPR, for example, would be programmed as F.007. G codes are sometimes used for switching from IPM to IPR or back again within the same program. For example, the G code to switch to IPM feed rate is G94 and the G code to switch to IPR is G95.

Some older CNCs used a format where the decimal point was not programmed but was fixed to allow for one place to the right of the decimal point. The rightmost number after the F represented the number after the decimal (for example, F5 represented 0.5 IPM, and F1000 represented 100 IPM).

Feed rate selection and use depend on many things, including the type of machine tool and cutting tool selected, the rigidity of the machine tool, the rigidity of the setup, and the material type. The maximum and minimum feed rates per axis of a machine tool will vary depending on the machine and control manufacturer. These feed rates dictate the permissible feed rate range.

Spindle speed, in revolutions per minute (RPM), is the number of revolutions that the spindle makes in one minute. The spindle speed code is an S word address followed by the rate in RPM. A spindle speed of 350 RPM, for example, would be programmed as S350. The spindle speed is generally governed by the work or cutter diameter, type of cutting tool, and material type. If the spindle speed is programmed in RPM, the spindle speed will remain constant until it is changed.

Spindle speeds in feet per minute (FPM)—sometimes referred to as constant surface speed (CSS)—are also used. Programming in FPM or CSS allows the machine to maintain a constant cutting speed regardless of the part diameter. For example, a facing pass programmed in CSS on a CNC turning center will cause the spindle to accelerate to keep the cutting speed constant as the cutting tool moves closer to the center of the part. If a succession of rough turns is taken, each pass will occur at a slightly higher spindle speed as the diameter gets smaller.

G codes are sometimes used for switching from RPM to CSS and back again within the same program. For example, switching to the CSS mode is accomplished with a G96 in this block:

N11 G96 S250 M3

Switching to the constant-RPM mode is done with a G97 in this block:

N13 G97 S1400 M3

Spindle speed ranges are extremely important to the success of a CNC machine installation. Like feed rate ranges, spindle speed ranges vary among machine and control manufacturers. But as long as programmers stay within the required spindle speed range, spindle speeds are infinitely variable. Additional consideration will be given to spindle speeds in the chapters on N/C turning and machining centers.

Programmable Z Depth

Most modern machine tools have programmable Z motion. Although the procedures differ in some aspects among manufacturers, the Z motion must be accurately programmed by the programmer if the machine tool is to produce quality workpieces.

Z coordinate input, like X and Y words in decimal point programming, uses only as many significant digits as required. The Z word address still needs to be programmed, but insignificant digits may be dropped. For example, if a Z depth of 2.5 in. is required, it would be programmed as Z−2.5. Some older CNCs used a seven-digit word preceded by a plus or minus sign and the Z word address.

The Z motion follows a rapid-to-position move, along the X- and Y-axes through either an R word (to be discussed later) or a rapid traverse fixed rate with a specific Z.

If only the Z word is being programmed to control the entire Z-axis movement, the following formula can be used in most cases:

$$Z = PS + CL + TL$$

where
- Z = distance from Z0 to spindle gage line
- PS = distance from Z0 (an arbitrary Z surface established by the programmer) to the part surface
- CL = clearance (if needed)
- TL = tool set length from spindle gage line to cutting edge*

To find the Z value for the part and tool in figure 5-6 (for which PS = 4 in., TL = 6 in., and CL = 0 in.), the following calculation is made:

$$Z = PS + CL + TL$$
$$Z = 4 + 0 + 6$$
$$Z = 10.0000 \text{ in.}$$

Additional applications of Z motion will be discussed later in a section on the R work plane as applied to fixed-cycle programmable Z motion.

---

*Since many systems are arranged with a tool-length storage feature, the control will in some cases add the tool set length (TL) to the programmed Z value. In such instances, the programmer need not include the tool set length dimension when calculating the Z word.

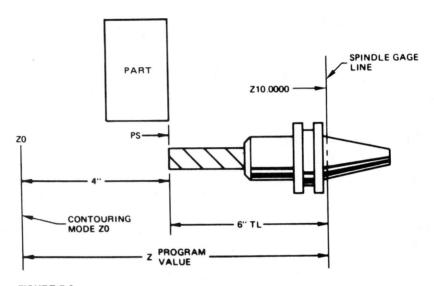

FIGURE 5-6
Part and tool for Z value calculation example.

Miscellaneous Functions

Miscellaneous functions perform a variety of auxiliary commands in numerical control. Generally, they are programmed through multiple-character, on/off codes that denote a function controlling the machine. These special codes, functional at the beginning or end of a cycle, are two-digit numbers preceded by the letter M (M**). They activate *auxiliary functions* such as spindle start, coolant control, and program stop.

The following is a list of explanations of basic miscellaneous functions in accordance with EIA standard coding:

- *M00 (program stop).* This code *inhibits* the reading cycle after the movement or function has been completed for the block in which the program stop was coded. In addition, this code also turns off the spindle and coolant if activated.

- *M01 (optional stop).* This code, like the M00 program stop, inhibits the reading cycle after the movement or function has been completed for the block in which the optional stop was coded. This code also turns off the spindle and coolant if activated. However, the code will function only if the operator has selected optional stop on the control. If no optional stop has been selected, the M01 will be read but no stop will occur.

- *M30 (end of program).* After the movement or function has been completed for the block in which the M30 end of program is coded, this code stops all *interpolation* (slide motion) and turns off the spindle and coolant. In addition, this code resets the programmed input to begin again or rewind the tape to the leader (front) portion of the tape.

- *M06 (tool change).* This function should be coded in the last block of information in which a given tool is used. The specific machine tool design determines the sequence of events during the tool change. This code also stops the spindle and coolant, if activated, and retracts the tool to the full retract position for the tool change.

Modal and Nonmodal Commands

Commands may be single, one-at-a-time commands, or they may stay in effect until changed by another command or canceled. Single, one-at-a-time commands are called *nonmodal* commands. A tool change (M06) is an example of a nonmodal command used to execute a single tool change at the end of an N/C block or instruction. Commands that stay in effect until changed or canceled are said to be *modal* commands. A feed rate (for example, F5.5) is an example of a modal command. The feed rate remains the same until another feed rate is selected later in the program.

SIMPLE PROGRAMMING EXAMPLES

The examples in this section illustrate and explain practical applications of some basic CNC functions, excluding part-holding methodology. They contain block-by-block explanations of each instruction to aid student understanding and comprehension.

The following part program is used for spot or center drilling holes 1 through 6 (see figure 5-7):

```
N10    T1 M6
N20    G0 X2 Y1.
N30    G81 R.1 Z−.5 F3.
N40    X4.
N50    X6.
N60    Y3.
N70    X4.
```

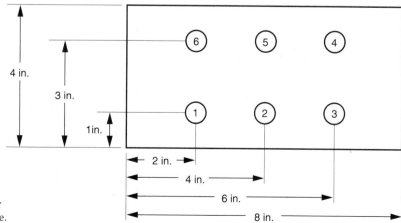

FIGURE 5-7
Sample part for drilling example.

N80	X2.
N90	G80 M9
N100	G0 G91 G28 Z0
N110	M30

G81 is a canned cycle used for drilling. G81 drills continuously down to the programmed Z depth on line N30, with no peck to break the chip. All canned cycles are modal, so the information on line N30 will be repeated at each location in lines N40 through N80. Let's look at each line.

N10	T1 M6	Tool change to tool 1.
N20	G0 X2. Y1.	Rapid traverse to the position of hole 1.
N30	G81 R.1 Z−.5 F3.	Continuous drill cycle. The value after the Z word is the total distance the tool travels from the Z zero point at the feed rate of 3 IPM. Once the tool reaches the Z depth, it rapid retracts back up to the R word's specified Z height.
N40	X4.	Canned-cycle repetition at position of hole 2.
N50	X6.	Canned-cycle repetition at position of hole 3.
N60	Y3.	Canned-cycle repetition at position of hole 4.
N70	X4.	Canned-cycle repetition at position of hole 5.
N80	X2.	Canned-cycle repetition at position of hole 6.
N90	G80 M9	G80 cancels the G81 canned cycle. M9 turns off the coolant.
N100	G0 G91 G28 Z0	Tool rapid traverses to its machine zero position.
N110	M30	Machine stops, all M codes are canceled, and program resets back to beginning.

A G83 canned drilling cycle can be added to the previous program to allow peck drilling and center drilling in the same program. G83 represents a peck drilling cycle, and the value after the Q word is the amount of peck. Other canned cycles could also be added to this program to perform such operations as tapping and boring. The revised and/or added code lines are as follows:

N110	G0 G91 G28 Z0	
N110	M6 T2	Tool change to tool 2.
N120	G0 X2. Y1.	Rapid traverse to position of hole 1.
N130	G83 R.1 Q.05 Z−.6 F3	Peck drilling cycle.
N140	X4.	Canned cycle repetition at position of hole 2.
N150	X6.	Canned cycle repetition at position of hole 3.
N160	Y3.	Canned cycle repetition at position of hole 4.
N170	X4.	Canned cycle repetition at position of hole 5.
N180	X2.	Canned cycle repetition at position of hole 6.
N180	G80 M9	G80 cancels the G83 canned cycle; M9 turns off the coolant.
N190	G0 G91 G28 Z0	Tool rapid traverses to its machine zero position.
N200	M30	Machine stops, all M codes are canceled, and program resets back to beginning.

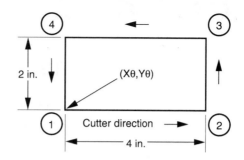

FIGURE 5-8

Sample part for simple milling example. 0.500-in.-diameter cutter

Consider a simple example in which a .500-in.-diameter cutter mills around the perimeter of a part (see figure 5-8). Notice that all the program moves take into account half the cutter diameter (that is, the radius). Code for this example would appear as follows:

```
N10   G0 X-.25 Y-.25      Rapid traverse to point 1
N20   G1 Z-.500 F20.      Feed down to a depth of -0.500 in.
N30   X4.250              Feed to point 2
N40   Y2.250              Feed to point 3
N50   X-.250             Feed to point 4
N60   Y-.250             Feed to point 1
```

REVIEW QUESTIONS

1. Name and describe briefly the three main parts of an N/C program.
2. What is the primary purpose of the sequence number in an N/C program?
3. How is the position of the decimal point accounted for when N/C coordinate information is coded in the word address/interchangeable format?
4. What important cutter-to-workpiece relationship must be considered when an N/C milling operation is being programmed?
5. In general, what is meant by a preparatory function?
6. What is the difference between canned cycle and noncanned cycle preparatory functions?
7. Name and describe some advantages of canned cycle preparatory functions.
8. What is the basic difference between the G81 and G82 codes?
9. What occurs each time the G84 tapping cycle is programmed?
10. Describe in detail the importance of programmed feed rates in an N/C program.
11. What are miscellaneous functions? What types of machine commands are controlled by these codes?

CHAPTER 6

Additional CNC Functions and Features

OBJECTIVES

After studying this chapter, you will be able to:

- Explain the value and use of polar coordinates
- Describe circular interpolation and its functional requirements
- Discuss cutter diameter compensation and tool offsets
- Describe how work surfaces and related changes are programmed
- Identify the importance of tool presetting and tool length compensation
- Identify the primary differences between random and sequential tooling
- Discuss adaptive machining and other CNC functions

POLAR COORDINATES

As we saw in Chapter 3, rectangular coordinates define the position of a point in two-dimensional space by the distance from each of two mutually perpendicular axes that intersect at an origin having zero value (three such axes are required for three-dimensional space). Polar coordinates define the position of a point in two-dimensional space by an angle and a radius from a fixed reference point or origin having zero value. Essentially, polar coordinates are used to define rotary and angular movements of part features, such as hole locations, using angles and radii.

In polar coordinates, the positive X-axis represents the *polar axis*. A line drawn from the origin to a designated point in the X-Y plane is the *radius*, or *vector*. The angle formed by the vector and the horizontal polar axis is the *polar angle*. The radius establishes the distance of the designated point from the origin, and the polar angle establishes the direction. Counterclockwise rotation from the polar line to the radius has a positive value, whereas clockwise rotation from the polar line to the radius has a negative value. For example, in figure 6-1, point A is 45° from the polar axis and has a radius of 1.25. Point B is 120°

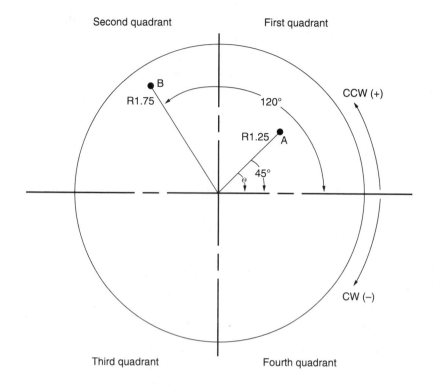

Polar coordinates: θ and R
Rectangular coordinates: X and Y
$\cos \theta = \frac{X}{R}$ and $\sin \theta = \frac{Y}{R}$
$\therefore$ X = R $\cos \theta$ and Y = R $\sin \theta$

FIGURE 6-1
Polar coordinates are used in programming when part features such as hole
locations are dimensioned in terms of angles and radii.

from the polar axis and has a radius of 1.75. Both are positive angles measured
counterclockwise from the polar axis.

Polar coordinates are sometimes used by designers to show angular relationships between two or more holes. However, the polar coordinates must be
properly referenced from the basic rectangular coordinates. The center point of
a bolt circle, for example, should be located in terms of the X and Y distances
from the workpiece zero point, and the circle's polar axis should be parallel with
the X-axis of the workpiece.

Parts that are dimensioned using angles and radii are always easier to
program using polar coordinates. A large percentage of the trigonometry calculations that programmers and operators make result from a need to convert
polar dimensions to rectangular coordinates. Fortunately today, many CNCs

will accept direct polar coordinate input such as bolt hole patterns and will automatically calculate the rectangular coordinates.

CIRCULAR INTERPOLATION

Circular interpolation allows the programmer to move the cutting tool in a circular path that can range from a small arc segment to a full 360° *span*. The cutter path along the arc is generated by the control system. On some CNCs, cutting a full 360° arc can be accomplished in one block of information. Older controllers required four arcs of 90° each to be programmed. The total number of blocks programmed will normally vary depending on the specific type of circular interpolation used.

In addition to the feed rate, there are four basic parameters necessary to program circular interpolation:

- Preparatory function (G02 or G03)
- Start point
- Center point
- End point

The preparatory function codes G02 and G03 are used for programming circular interpolation. These codes determine the *direction* of the circular path as viewed from the positive end of the axis that is perpendicular to the plane of interpolation. G02 represents clockwise (CW) circular interpolation, and G03 represents counterclockwise (CCW) circular interpolation. These codes are programmed in the block where circular interpolation becomes effective. They remain effective until a new preparatory function code is programmed.

The *start point*, in the XY, YZ, or ZX plane, is usually the end point of a previous arc (circular interpolation) or the end point of a line (linear interpolation). The start point is always described by X, Y, and/or Z words, and it normally positions the cutting tool for the following circular move.

The *center point* (in the XY, YZ, or ZX plane) is the center of the circular arc. The center point is described by I, J, and K words. The I word describes the X coordinate value, the J word describes the Y coordinate value, and the K word describes the Z coordinate value. If the control knows the start point and center point, it knows the radius. On some controllers, the radius for the arc can be substituted for the center.

The *end point* can be expressed in terms of Cartesian coordinates or polar coordinates on many controls. If polar coordinates are used, the end point will be specified as the number of degrees through which to rotate, rather than as an XY or XZ distance.

Some controllers expect the circular interpolation I and J words to be the incremental distances from the arc start point to the arc center point. Such controllers assume I and J are incremental even if the program is absolute. Other controllers assume the I and J words are used to specify the absolute location of the arc center point. During programming, it is extremely important to know precisely how circular interpolation is handled because mistakes can be very dangerous and costly. Always consult the specific operator/programmer's manual before attempting to program any CNC machine tool.

The following two examples demonstrate circular interpolation. Both examples illustrate the two most commonly used methods of programming circular interpolation today: the center point method and the radius method.

Clockwise Circular Interpolation for a 90° Arc

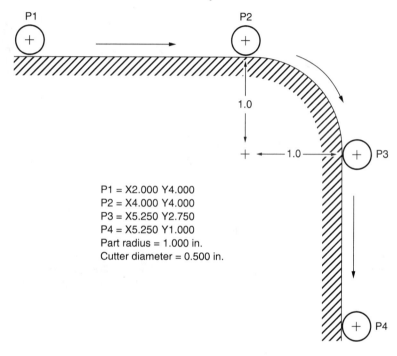

P1 = X2.000 Y4.000
P2 = X4.000 Y4.000
P3 = X5.250 Y2.750
P4 = X5.250 Y1.000
Part radius = 1.000 in.
Cutter diameter = 0.500 in.

CENTER POINT METHOD

```
N20  G01  X2     Y4   F6              Move to P1 at 6 IPM
N21  X4                                Move to P2 (start point)
N22  G02  X5.25  Y2.75  I4   J2.75     C.I from P2 to P3 (end point) CW
N23  G01  Y1                           Move to P4
```

RADIUS METHOD

```
N20  G01  X2     Y4   F6              Move to P1 at 6 IPM
N21  X4                                Move to P2 (start point)
N22  G02  X5.25  Y2.75  R1.25          C.I. from P2 to P3 (end point) CW
N23  G01  Y1                           Move to P4
```

Counterclockwise Circular Interpolation for a 180° Arc

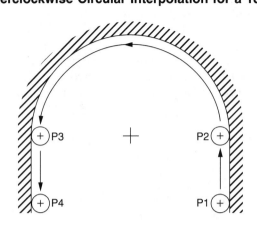

P1 = X1.75 Y−1.5
P2 = X1.75 Y0
P3 = X−1.75 Y0
P4 = X−1.75 Y−1.5
Part radius = 2.000 in.
Cutter diameter = 0.500 in.
Arc center = 0.0

CENTER POINT METHOD

N30	G01	X1.75	Y−1.5	F6		Move to P1 at 6 IPM
N31	Y0					Move to P2 (start point)
N32	G03	X−1.75	Y0	I0	J0	C.I. from P2 to P3 (end point) CCW
N33	G01	Y−1.5				Move to P4

RADIUS METHOD

N30	G01	X1.75	Y−1.5	F6	Move to P1 at 6 IPM
N31	Y0				Move to P2 (start point)
N32	G03	X−1.75	Y0	R1.75	C.I. from P2 to P3 (end point) CCW
N33	G01	Y−1.5			Move to P4

There are other types of interpolation available with modern MCUs, including parabolic, cubic, and helical interpolation. *Parabolic interpolation* is used to approximate curved sections that conform to either a complete parabola or a portion of one. Cubic interpolation is applicable to automotive shapes requiring third-degree curve interpolation of sheet metal–forming dies. Helical interpolation lends itself to helical cutting applications where the control must calculate the radius of the helix from the start of the arc to the center. If the end point does not fall on the arc defined, the control will interpolate a helical arc as far as possible and then move to the programmed end point with a linear move.

Parabolic, cubic, and helical interpolation are specialized applications for the particular needs of industries that manufacture components with complex shapes. The most common interpolation routine is circular. It lends itself to a variety of common manufacturing applications.

TOOL OFFSETS

An *offset* is a value that the control reads and stores as the radius or length of a cutting tool. Offset values are manually entered in the MCU to make adjustments to the cutting path based on the length and radius of the specific cutter being used. Tool length offsets and tool radius offsets are stored in the control but outside the program. Offsets define the cutter geometry and position for the CNC's computer to use in determining the cutter path. Tool radius offsets can work in conjunction with cutter diameter compensation (explained below) by telling the control that a particular cutter is either undersize or oversize relative to the diameter originally intended in the program.

Offsets are entered during the setup and can be manually changed to suit available cutter sizes and cutting conditions. When a cutter is changed, the offset is reentered according to the new parameters. The CNC's computer reads these new parameters and adjusts the cutter path accordingly.

Offsets have four basic uses:

1. They allow a "close-to" diameter cutter to be used when the originally intended cutter diameter is unavailable (explained in detail later).

2. Entering an offset value that doesn't match the cutter being used allows one to "trick" the control into cutting the part either undersize or oversize (see figure 6-2). The original program is essentially altered through the use of a different cutter diameter along the program's original path.

3. They can be used to compensate for tool wear. As a tool dulls, the offset can be changed to "trick" the control into moving the cutter slightly closer to the cutting surface. However, this is not a good practice because surface finish and other problems may arise from using dull tools. Dull tools should be replaced with sharp cutting tools.

4. Through the use of two different sets of tool offsets, rough and finish passes may be made using the same program and the same cutting tool. The first offset tells the control that the cutter is larger than it really is, which causes the control to back the cutter away from the part, thereby leaving finish stock. Recutting the part using the second offset—which represents the correct cutter diameter—will then result in a part with the correct finish size.

In the control, offset values are stored in the TOOL OFFSET registers. The length offset of a particular tool is stored in the register location corresponding to the pocket location where the tool itself is stored. For example, the length offset for the tool in pocket location 12 would be stored as entry 12 in the TOOL

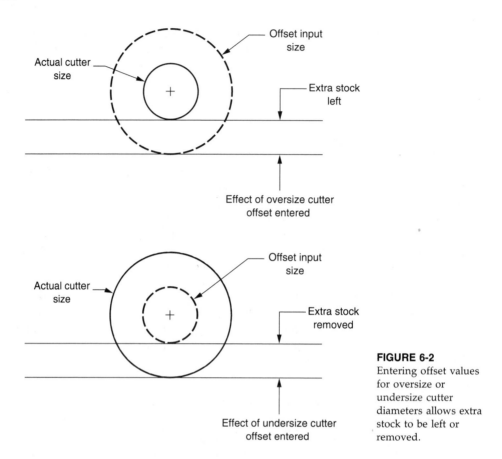

FIGURE 6-2
Entering offset values for oversize or undersize cutter diameters allows extra stock to be left or removed.

OFFSET register. The radius offset location for a tool on some controls is found by adding 50 to the length offset location (location 62 for a length offset location of 12).

CUTTER DIAMETER COMPENSATION

Most builders of numerical control units offer some type of *cutter diameter compensation* (CDC). Cutter diameter compensation enhances the machine and control system by providing the capability to do the following:

1. Control the size of milled close-tolerance slots, pockets, bores, and steps
2. Use a cutter of a different diameter than that originally intended when the part was programmed

CDC permits oversize or undersize cutters to be used while still maintaining the programmed part geometry. The difference between the cutter diameter programmed and the one used—known as a *CDC value*—generally ranges from −1.0000 in. to +1.0000 in. Methods of inputting and compensating for cutter size differences vary among control manufacturers.

When cutter diameter compensation is being used, a CDC value can be entered into the control for each tool number programmed. Inputting a CDC value for one tool does not affect CDC values for other tools. The CDC value becomes active when the appropriate tool is loaded into the spindle. If the value is changed after the tool is loaded, it becomes active on the next span prepared by the control system. CDC, effective only when the appropriate codes are programmed, can be programmed for both linear and circular interpolation.

When programming a workpiece for any milling operation, the operator/programmer must take three main factors into consideration:

1. The coordinates/dimensions of the surfaces to be cut.

2. The cutter direction and position relative to the surface or programmed line to be machined. (When CDC is *not* in effect, the cutter's position is already fixed by coordinate locations in the program.)

3. The radius of the cutting tool.

If CDC will be used, specific codes must be added. The G codes used for programming CDC are:

- G40—cancel cutter compensation code for G41 or G42.

- G41—cutter compensation left of line. This code directs the cutter to the left side of the programmed cutting line.

- G42—cutter compensation right of line. This code directs the cutter to the right side of the programmed cutting line.

When deciding which code to use (G41 or G42), the operator/programmer must consider the direction of the cutter path and the proper cutter position. After determining the direction of the cutter movement, he or she must visualize "walking directly behind" the cutter. This determines the cutter position relative to the drive surface or cutting line. Imagine following directly behind cutter A in figure 6-3.

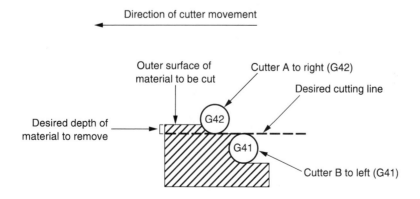

FIGURE 6-3
G42 directs the cutter to the right side of programmed line based on cutter direction.

The cutter is positioned on the right side of the desired cutting line. G42 is the proper code for this arrangement. If G41 were selected for this direction of cutter movement, the cutter would position itself on the opposite side of the desired cutting line (cutter B), resulting in too much material being removed, thereby risking possible tool breakage, machine damage, part scrapping, and potential operator injury.

In figure 6-4, the direction of the cutter movement has changed. Again, visualize walking behind cutter A and following the direction of the cutter movement. The G41 code (left) places the cutter on the proper side of the desired cutting line for this programmed move. Selecting the G42 code in this example would have the potentially dangerous results mentioned above.

After the proper cutter compensation code has been selected, a D statement must directly follow, as the following code example demonstrates:

G01 G41 D62 X1.5 Y0 F20.

The number 62 after the D is the assigned location in the control's TOOL OFFSET register where the tool radius offset is stored. If the tool has a 1.0-in. diameter cutter, the tool radius offset value entered in location 62 of TOOL OFFSET will be 0.500.

Rules governing cutter diameter compensation are as follows:

1. A D statement must follow a G41 or G42 programmed code.

2. Cutter diameter compensation should be canceled at the end of a program.

3. When the radius of a corner is smaller than the cutter radius, an alarm is normally generated, and the CNC will stop at the start of the block.

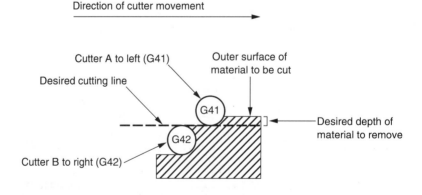

FIGURE 6-4

G41 directs the cutter to the left side of the programmed line based on cutter direction.

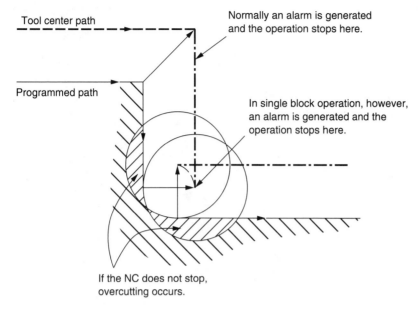

Tool center path

Normally an alarm is generated and the operation stops here.

Programmed path

In single block operation, however, an alarm is generated and the operation stops here.

If the NC does not stop, overcutting occurs.

FIGURE 6-5
Case in which the radius of a corner is smaller than the cutter radius.

In single-block operation, however, overcutting will take place because the tool doesn't stop until after the block has executed. This condition is illustrated in figure 6-5.

Machining a groove smaller than the tool diameter would force the path of the tool to move in the reverse of the programmed direction. This condition is illustrated in figure 6-6. In this situation, overcutting would result, so an alarm will be generated at the start of the block.

4. Codes G02 and G03 are not allowed in the startup block for CDC. For example,

G41 G02 X3.5 Y0 R.5

is invalid. If this block is programmed, the machine will enter an alarm state.

5. The program should not call for two consecutive blocks without tool movement when cutter diameter compensation is in effect. If it does so, tool movement equal to the amount of the tool radius offset will be produced, or an alarm state may result.

Other conditions and examples relating to cutter diameter compensation are covered in the CNC manufacturer's operator and programming manuals. These manuals are specific to each machine and control type and should always be consulted before attempting to program any CNC.

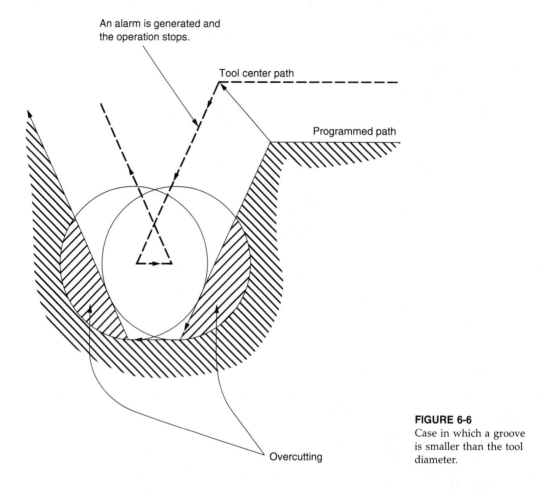

An alarm is generated and
the operation stops.

Tool center path

Programmed path

Overcutting

FIGURE 6-6
Case in which a groove
is smaller than the tool
diameter.

TOOL LENGTH COMPENSATION

Allowances must be made for differences in tool lengths because they may vary considerably from one cutting tool to the next. Most companies overcome these differences in tool length by having all tools preset. Presetting tools (setting or assembling to a predetermined length before use) establishes consistency of length each time a particular tool is required. Tool length values are then entered into the MCU.

Some companies have prepared *tool assembly drawings* for each tool. These tool layouts are prepared to describe the cutting tool and its setting, or preset, length (see figure 6-7). Each tool is represented by a tool assembly number. On machines and controls equipped with tool assembly number storage, this is the number accessed on the machining center and stored within the control. For example, in figure 6-7, tool assembly number 14031416 has a stored length value of 6.75 in.

The standard center drill for the tool assembly shown in figure 6-7 must be set so that its dimensions are consistent with those given on the drawing.

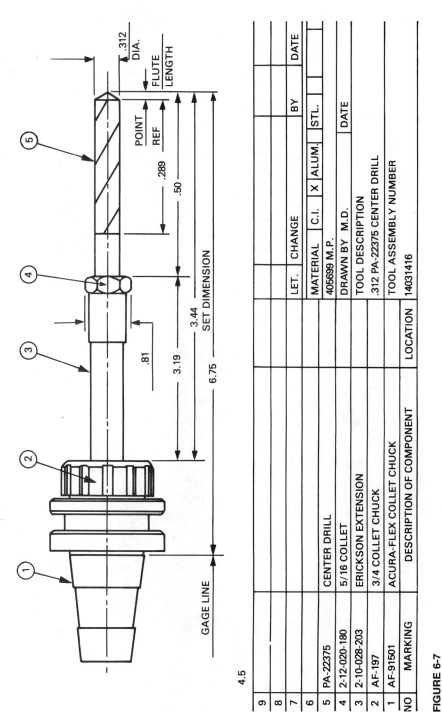

NO	MARKING	DESCRIPTION OF COMPONENT						LOCATION
9								
8								
7								
6	PA-22375	CENTER DRILL			LET.	CHANGE		BY	DATE
5	2-12-020-180	5/16 COLLET						
4	2-10-028-203	ERICKSON EXTENSION		MATERIAL	C.I.	X	ALUM.	STL.
3	AF-197	3/4 COLLET CHUCK		405699 M.P.				
2	AF-91501	ACURA-FLEX COLLET CHUCK		DRAWN BY M.D.				DATE
1				TOOL DESCRIPTION				
				.312 PA-22375 CENTER DRILL				
				TOOL ASSEMBLY NUMBER				
				14031416				

FIGURE 6-7

A typical tool assembly drawing.

Generally, the machine operator or individuals in a tool store or tool-preset area will assemble the drill in its holder according to the tool assembly drawing each time the tool is required. Preset and tool store personnel will sometimes maintain a duplicate set of drawings consistent with tools selected for use in N/C programming. When the part is ready to be processed, the preset area assembles all the tools specified by the part programmer according to each tool assembly number.

In some cases, an electronic or optical tool gage is used to obtain the overall length of the tool. Tool-presetting gages are based on touch-readout tool gages (see figure 6-8) or optical projection systems that magnify the tool point (see figure 6-9). Recent advances in tool presetting have allowed tool-presetting gages to interface directly with the MCU for direct tool length entry into the control system. This type of gaging and machine control interface helps obtain accurate tool lengths and reduce human data entry error.

It is important to remember that the machine must be made to cut metal so as to maximize productivity as much as possible. Therefore, arrangements must be made to replace dull cutting tools in a minimum amount of time and still maintain accurate tool lengths.

FIGURE 6-8
An electronic tool-presetting gage. (Courtesy of Cincinnati Milacron Inc.)

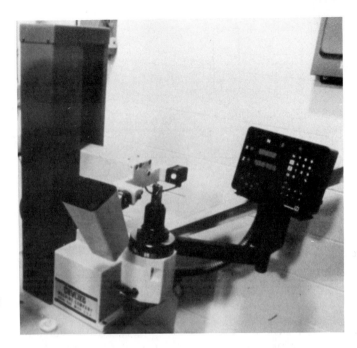

FIGURE 6-9
An optical tool-presetting gage. (Courtesy of Cincinnati Milacron Inc.)

R WORK PLANES

In an N/C program, the R word, first introduced in Chapter 5, establishes a "rapid to" location for a workpiece just above the work surface.

The *R work planes*, sometimes called *Z rapid position planes*, are parallel to the work surfaces and are slightly above them (by the amount specified in the R word). A work plane establishes a reference location at a specified height above a particular work surface. The original R work plane is called the R0 work plane. All other work planes are given relative to the R0 plane, thereby establishing height or distance for the workpiece. On some N/C machines, the R0 work plane is defined relative to the highest part surface.

As figure 6-10 shows, R0 is sometimes established above the highest work surface, and all tools are set at this surface. When setting up the job, the operator might place a 0.100-in. gage block on top of the highest work surface (or whatever surface the programmer has specified).

The turret is lowered, or the spindle is brought down to the workpiece, by jog control, so that the tool in the spindle just touches this 0.100-in. gage block. The operator then sets tool length compensation to R0, and the rapid distance above the part surface for that tool is complete. This same process must be repeated for all tools because they are all of different lengths. Generally, it is not necessary for one to add in the gage height distance when changing work surfaces, as figure 6-10 shows. In the case of some control manufacturers, the gage height distance of 0.100 in. is built into the MCU. Consequently, whenever a Z feed motion is called for, the 0.100 in. will be automatically added to the programmed Z depth.

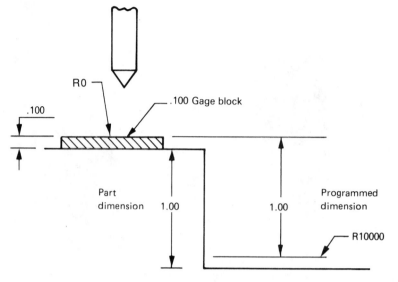

FIGURE 6-10
Establishing the R work plane.

When programming the R word, and particularly when changing work surfaces, the programmer must always know the location of each tool's tip. When changing work surfaces, the programmer simply programs the R distance indicated on the blueprint. As shown in figure 6-10, a rapid traverse distance of 1 in. is specified as an R1.0000; in this case, it positions the tool 0.100 in. above the part surface.

A Z depth of −.375 (Z −.375) programmed at R0, for example, could also be programmed at R1.0000. In both instances, the tool will go 0.375 in. into the workpiece relative to the respective work planes. In addition, a programmed R is modal. That is, the programmed R word stays locked in the R register until a new R word is programmed. When a new R word is programmed, a change in work surface will result.

Whether the programmer wants to move the tool to a higher or lower work surface is of considerable importance. Machine tools that require an R word to be programmed usually have a specific order in which blocks of information are processed. For a block of information that contains an X, a Y, an R, and a Z word, the X and Y words have the highest priority and will be satisfied first. The R word has the second highest priority and will be satisfied second. The machine tool will respond with a rapid traverse to satisfy these words at the rapid traverse rate of the machine tool (400 to 1200 IPM or higher). The Z word has the lowest priority; the tool will feed to depth only after the X and Y words have positioned the cutting tool and the R word has been established for the programmed work surface.

No tool-collision problems occur during a change from a higher to lower work surface (see figure 6-11) because of the order in which the programmed words are followed (X and Y first, R second, and Z last). For a change from a lower work surface to a higher one, the priorities are the same but must

| O/N | G | X± | Y± | Z± | R± | I/J/K | A/B/C | F/E | S | D/H | T | M |
SEQ.	PREP. FUNCT.	POSITION	POSITION	POSITION	POSITION	POSITION	P/Q ± WORD	FEED RATE	SPINDLE SPEED	WORD	TOOL WORD	MISC. FUNCT.
Ø 5	G81	X+ 80000	Y+100000	Z-10000	R+100000		B 0	F 150	S 720		T3	M03
N 6			Y+140000		R 90000							

— In Sequence Number Ø5, the G81 code rapid advances the X- and/or Y-axes simultaneously to Pos. 1 from the previous position. When Pos. 1 is reached, the Z-axis will rapid to the R 10.0000 plane, and will feed to the programmed depth of 1'' at the programmed rate. After reaching depth, the Z-axis will rapid retract to the R 10.0000 plane, and the next block of information will be read and acted upon.

— In Sequence N6, the G81 code rapid advances the Y-axis — at the R 10.0000 plane — to Pos. 2. Then the Z-axis will rapid to the new R plane (R 9.0000), and feed to the programmed depth of 1''. After reaching depth, the Z-axis will rapid retract to the R 9.0000 plane.

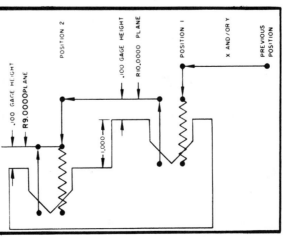

FIGURE 6-11
Changing the work plane (high to low).

O/N SEQ.	G PREP. FUNCT.	X± POSITION	Y± POSITION	Z± POSITION	R± POSITION	I/J/K POSITION	A/B/C P/Q ± WORD POSITION	F/E FEED RATE	S SPINDLE SPEED	D/H WORD	T TOOL WORD	M MISC. FUNCT.
4	G81	X+ 80000	Y+ 90000	Z- 10000	R+ 90000		B 0	F 150	S 720		T 3	M03
5	G80				R 100000							
6	G81		Y 60000									

— In Sequence Number Ø4, the G81 code rapid advances the X- and/or Y-axes simultaneously to Pos. 1 from the previous position. When Pos. 1 is reached, the Z-axis will rapid to the R 9.0000 plane, and will feed to the programmed depth of 1" at the programmed rate. After reaching depth, the Z-axis will retract to the R 9.0000 plane, and the next block of information will be read and acted upon.

— Sequence Number N5, with the G80 code, will rapid retract the tool from the R 9.0000 plane to the R 10.0000 plane.

— Sequence No. N6, with the G81 code re-programmed, will rapid advance the Y-axis to Pos. 2 and will drill a hole 1" deep on the higher level.

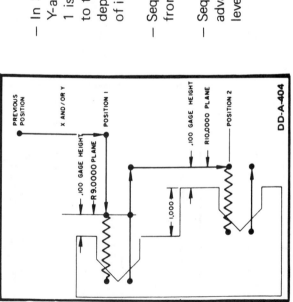

DD-A-404

FIGURE 6-12
Changing the work plane (low to high).

be programmed differently. For a change from a lower to higher work surface, programming a block of information with X, Y, R, and Z in the same line of information would result in a collision of the tool at rapid traverse against the side of the workpiece. This could cause tool breakage, injury to the operator, and considerable damage to the toolholder and machine spindle, as well as other serious effects. This problem must be overcome by first retracting the tool to the higher work surface with a G00 (G80 on some older CNCs) and then reinstating the desired code plus the new X and Y positions, as shown in figure 6-12. Being aware of clamp locations is also important during programming. The programmer may find it necessary to avoid accidents by programming Z avoidances in order to move over or around clamps.

When preparing a program, the programmer must know what surface the programmed tool is on, where the tool will rapid traverse to next, and the path the tool will follow to reach the next position. If programmers overlook potential collisions when checking their manuscripts, critical and expensive accidents can occur.

It should be noted that each machine tool and control manufacturer has its own programming specifications and requirements for its equipment. The correct programming manual should always be consulted before attempting to program a specific machine tool and control system.

ADAPTIVE CONTROL

When programming N/C equipment, the programmer usually determines the feeds and speeds based on the tool type and diameter, material type, setup rigidity, etc. Usually, optimum feeds and speeds are chosen so as to make the N/C machine and cutting tool as productive as possible. However, ideal productivity does not occur often because of excessive material hardness and dull cutting tools. Tool breakage may occur until the feeds and speeds are cut back to accommodate the particular machining circumstances.

Adaptive control, or torque-control machining, was developed to speed up or slow down a cutting tool while the tool is engaged in the actual cutting operation. The function of adaptive control is to sense machining conditions (see figure 6-13) and adjust the feeds and speeds accordingly. Sensing devices are built into the machine spindle to sense torque, heat, and vibration. These sensors provide feedback signals to the MCU, which contains the preprogrammed safe limits. If the preprogrammed safe limits are exceeded, the MCU alters or adjusts the feeds and speeds.

Programming requirements are basically the same as usual for adaptive control. However, it may be necessary to insert specific codes in order to turn the function on or off. The types of adaptive control and how they are used will vary, of course, among machine and control manufacturers.

The use of adaptive control is becoming increasingly popular as more companies try to optimize machine spindle time and reduce tooling requirements. Adaptive control provides automatic optimization of N/C machining operations for part-manufacturing facilities.

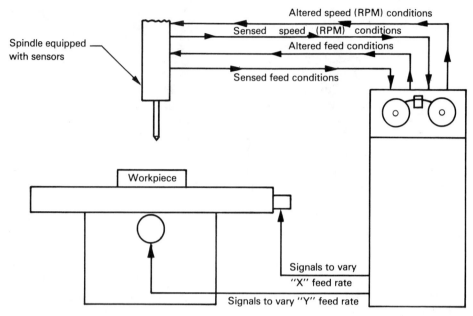

FIGURE 6-13 •
Sensory and feedback adaptive control signals.

RANDOM AND SEQUENTIAL TOOLING

Both random and sequential tooling are used on tool-changing machine tools. *Random* refers to a lack of a rigid pattern of tool selection. *Sequential* refers to tools that are accessed in a particular order. Sequential tool selection requires the tools to be loaded in the exact order they will be used in the program. When the program begins, the tools are selected and used one after the other, maintaining the established sequence. The correct sequential loading of the tools is of primary importance to the operator for the successful execution of the part program. If, for some reason, the tools are placed out of order, the N/C machine will not know the difference; it will just change to the next tool in sequence and place it in the spindle. Consequently, it may try to drill with a tap. This may cause significant injury to the operator, as well as damage to tools and equipment. In addition, once a tool takes its turn in sequence and is returned to the *tool storage drum*, it cannot be used again. The sequence would have to be broken for a particular tool to be reused. Normally, if a tool of the same diameter is required more than once in a program, two tools of the same length and diameter are programmed in their proper sequence. Sequential tooling was used on some of the earliest machining centers and is for the most part obsolete today.

Random tool selection is common in industry today because of the versatility it provides over sequential selection. When a tool change is called for, the tool changer arm removes the previous tool, puts the next tool in the spindle, and places the previous tool back in the tool magazine (in the specific pocket assigned to that tool assembly number). The CNC remembers the location of the tool by means of the tool assembly number initially input to the control, and it assigns a pocket location. The important feature of random tooling is that any tool can be accessed by the MCU and loaded into the machine spindle at any time. The MCU does not care about the order of the tools or whether they have previously been used.

The programmer must be especially careful to ensure that the workpiece is moved far enough away from the spindle before a tool change is executed for either random or sequential tooling. Such precautions help prevent collisions from occurring when a tool is being loaded into the spindle. The programmer will often direct the machine tool to its full retracted position prior to any tool changes.

OTHER FUNCTIONS

Many machine and control options exist that provide increased technical capabilities for both programming and operation of CNC equipment. Although it is impossible to list and discuss all options, the following subsections introduce additional functions and features on a variety of modern equipment.

Mirror Image

The *mirror image* function enables an opposite, or left-hand, part to be made from the same program. The part produced will be a mirror image of the actual part. The CNC accomplishes mirror imaging by switching +X moves to −X moves, for example, or by switching +Y moves to −Y moves. Mirror imaging offers users the ability to produce a second, left-hand version of the original part without reprogramming. Mirror imaging may be offered as a standard feature on some CNCs and as an option on others.

Scaling

Scaling offers the programmer or operator the ability to expand or contract the entire part being programmed. The scaled part is defined in terms of a percentage of the original programmed moves. For example, scaling by 75% would produce CNC movements of 0.75 in. for every inch of movement in the original program. Scaling by 125% would produce CNC movements of 1.25 in. for every inch of movement in the original program. The main application of the scaling function is in plastic moldmaking, where shrink factors must be built into the mold and the mold must be bigger than the finished part to allow for the shrinkage.

Tool Trim

The *tool trim* function permits the operator to adjust the Z-axis command positions to compensate for inaccuracies that could result in variations of machining depths. Tool trim codes are usually two-digit words preceded by a letter (sometimes D). The code specifies the trim value from a group, that is to be operative during a portion of the program. The trim value is entered into the CNC by the operator. The values normally range from ±0.0001 to ±1.0000 in. Negative values move the tool tip closer to the work surface, and positive values move it away from the surface. Tool trims for a particular tool stay in effect until canceled by a new tool trim code, tool change, end of program, or data reset.

Tool Usage Monitor

The tool usage monitor keeps track of the actual time of tool use versus the predicted effective tool life for the particular tool. The predicted tool life can be entered by the operator via the CNC keyboard. If the tool cycle time expires while the tool is in the spindle, an error message will be displayed on the CRT. The machining cycle will not be inhibited, but the operator will be notified that the tool should be replaced.

Tool Setup Identification

The *tool setup identification* feature allows tooling to be assigned to a specific setup. Tools may also be shared between setups, and common tools may be left resident in the tool storage mechanism. When a new setup is being placed on the machine or when tooling for dual fixturing or pallet shuttle machines is being coordinated, the CRT displays which tools are to be added or deleted during job setup and removal.

EXAMPLE PROGRAMS

The following examples illustrate the use of cutter diameter compensation and circular interpolation, excluding part holding and clamping.

Consider first a straight periphery milling case in which the cutter is offset with cutter diameter compensation (see figure 6-14). To eliminate accounting for the cutter diameter in each move, cutter diameter compensation codes G41 and G42 can be used to automatically adjust for the cutter diameter offset. *This allows actual part dimensions to be programmed.*

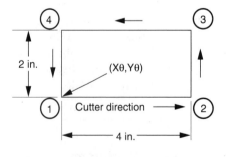

0.500-in.-diameter cutter

FIGURE 6-14
Part diagram for straight
periphery-milling example.

G41 corresponds to being to the left of the line in which the cutter is traveling, and G42 corresponds to being to the right.

The program to achieve this milling operation would appear as follows:

```
N001   G0 X-.300 Y-.300
N002   G1 Z-.250 F20.
N003   G1 G42 D51 X0 Y0 F20.
N004   X4.00
N005   Y2.00
N006   X0
N007   Y0
```

Let's consider each line individually.

N001	G0	X−.300 Y−.300	Machine rapid traverses to position outside of first corner to be milled. It is common practice to begin from a point that is more than a cutter radius away from the start point (in this case, the radius is 0.25 in., so the starting coordinates fit this criterion).
N002	G1	Z-.250 F20.0	Machine feeds down to a depth of 0.25 in. in the −Z direction at a feed rate of 20 IPM.
N003	G1	G42 D51 X0 Y0 F20.	Cutter moves to the start point (point 1) at (X0, Y0), with cutter diameter compensation to the right activated by G42. D51 specifies the location in the tool library where the control learns the size of the cutter radius. F20. sets up a feed rate of 20 IPM
N004	X4.00		Cutter moves to point 2 at previous feed rate.
N005	Y2.00		Cutter moves to point 3.
N006	X0		Cutter moves to point 4.
N007	Y0		Cutter moves to point 1.

Next, consider an example of straight and circular periphery milling with cutter diameter compensation. The dimensions are shown in figure 6-15.

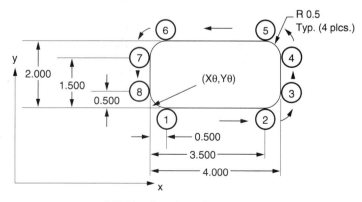

FIGURE 6-15
Part diagram for straight and circular periphery-milling example.

0.500-in.-diameter cutter

The code to perform this operation would appear as follows:

```
N005   G0 X-.500
N010   G1 Z-.250 F20.
N015   G1 G42 D51 X0 Y0
N020   X3.500
N025   G3 X4.000 Y.500 R.500
N030   G1 Y1.500
N035   G3 X3.500 Y2.000 I3.500 J1.500
N040   G1 X.500
N045   G3 X0 Y1.500 I.500 J1.500
N050   G1 Y.500
N055   G3 X.500 Y0 R.500
```

Let's consider the individual code lines.

N005	G0 X−.500 Y−.500	Machine rapid traverses to start point (−0.5, −0.5).
N010	G1 Z−.250 F20.	Machine feeds down to a depth of 0.25 in. in the −Z direction at a feed rate of 20 IPM.
N015	G1 G42 D51 X0 Y0	Machine turns on cutter diameter compensation to the right (G42), reads preset diameter of tool (D51), and moves to XY coordinate (0,0).
N020	X3.500	Machine moves to position 2.
N025	G3 X4.000 Y.500 R.500	Machine moves in a circular, counterclockwise direction (G3), around a 0.5-in. radius, to end points X and Y (point 3).
N030	G1 Y1.500	Machine moves in a linear direction to point 4. (*Note:* G3 is modal, so G01 must be reprogrammed for the linear move.)
N035	G3 X3.500 Y2.000 I3.500 J1.500	Machine moves in a circular, counterclockwise direction (G3) to end points X and Y (point 5) around a center point location given by I (for X) and J (for Y).
N040	G1 X.500	Machine moves in a linear direction to point 6.
N045	G3 X0 Y1.500 I.500 J1.500	Machine moves in a circular, counterclockwise direction (G3) to end points X and Y (point 7) around a center point of I and J.
N050	G1 Y.500	Machine moves in a linear direction to point 8.
N055	G3 X.500 Y0 R.500	Machine moves in a circular, counterclockwise direction (G3), around a 0.5-in. radius, to end points X and Y (point 1).

Note: In this example, two corners were milled using the radius method of circular interpolation and two corners were milled using I + J words for circular interpolation. This was done to demonstrate, in one example, the two different methods of programming circular interpolation that we have learned.

REVIEW QUESTIONS

1. Explain what is meant by polar coordinates, and describe their value to operators and programmers.

2. What is circular interpolation? What four parameters must be programmed in circular interpolation?

3. Name other types of interpolation available on modern MCUs, and briefly discuss what they are used for.

4. What are the two primary reasons for using cutter diameter compensation?

5. Name the three uses of tool offsets.

6. What is the order of processing for a block of information (in the word address, interchangeable, or compatible format) that contains X, Y, R, and Z word information? What types of problems may arise during a change from a lower work surface to a higher one? How can these problems be avoided in a program?

7. What are preset tools, and why are they important? What consistency is achieved by maintaining a library of tool assemblies?

8. Explain the difference between sequential and random tooling. Which is most widely used today? Why?

9. Describe adaptive control and its primary purpose. What advantages can be gained from adaptive control?

10. Name some other CNC features that exist on modern N/C equipment, and briefly explain their functions.

CHAPTER 7

CNC Turning Centers and Programming

OBJECTIVES _____

After studying this chapter, you will be able to

- Identify turning center types and describe principal design attributes and related advantages
- Explain primary turning center axes, programming, principles, and format information
- Identify the types of operations performed on a CNC turning center
- Explain modern work-holding and work-changing equipment and methods
- Describe system subroutines and their primary importance
- Discuss new automated features and capabilities

The CNC lathe, which only partially resembles manual models, continues to be the mainstay of rotational-part metal-removal machine tools. However, CNC lathes today must not only be able to handle increasing demands for flexibility, automation, and higher horsepower, speeds, and stock removal rates, but must also permit work-area access by such devices as robot loaders/unloaders, chuck-changers, and tool-changers.

Now commonly referred to as *turning centers* because of their increased flexibility and capability, CNC lathes are classified as one of two types: vertical and horizontal. Vertical CNC turning centers (see figure 7-1) are modern adaptations of the manual vertical turret lathes (VTLs). Although principally designed for larger and unwieldy workpieces, vertical turning centers have advanced considerably in terms of state-of-the-art CNC features and technology. Except for the enhanced features and added technology, though, the basic construction of vertical turning centers has been, for the most part, essentially unchanged.

Horizontal CNC turning centers of the shaft, chucker, or universal type (see figures 7-2 and 7-3) have changed not only in terms of advanced features and technology, but in basic construction as well. Many modern horizontal CNC turning centers are of the slant-bed design (see figure 7-4), some with outside-diameter (OD) and inside-diameter (ID) tools mounted on the same indexable turret. Advantages of the slant-bed design include easy access for loading,

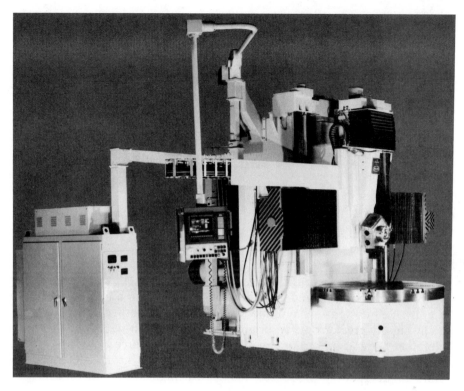

FIGURE 7-1
Vertical CNC turning center. (Courtesy of DeVlieg Sundstrand)

FIGURE 7-2
Horizontal CNC turning center. (Courtesy of Mazak Corp.)

FIGURE 7-3
Horizontal CNC turning centers come in a variety of types and sizes.
(Courtesy of Mazak Corp.)

unloading, and measuring; allowance for chips to fall free; minimum floor-space utilization; ease and quickness of tool changes; and better strength and rigidity. Bed designs are typically constructed of steel weldments, which generally supply greater rigidity, or are made of cast iron, which tends to dampen vibration better. Some beds are even made of reinforced concrete and epoxy-granite resin. Machine tool beds composed of this material provide excellent vibration-dampening properties, along with high static and dynamic rigidity.

FIGURE 7-4
Horizontal CNC turning center of slant-bed design. (Courtesy of Cincinnati Milacron Inc.)

Development continues on the use of various materials, such as steel-concrete composites and ceramics, for machine tool bed and base construction.

Increased performance requirements have brought about considerable change in turning center headstock and drive systems. Headstocks are either gear driven or gearless (belt or motorized direct drive). Geared types typically have broader speed ranges, accommodating the needs of different materials that require constant cutting speeds from low to high ranges. Gearless drives, which have become increasingly popular, offer the advantages of high speed and improved surface finish and accuracy as a result of eliminated gears, clutches, and shafts. In addition the elimination of these parts has resulted in lower costs. However, the choice of drive depends on the application, and the user must weigh such variables as workpiece size and material to determine which type best suits overall requirements.

CNC LATHE AXES

A basic CNC turning center (see figure 7-5) uses only two axes: Z and X. The Z-axis is taken to be a line drawn through the center of the machine spindle. For both OD and ID operations, a negative Z (−Z) is a movement of the saddle toward the headstock. A positive Z (+Z) is a movement of the saddle away from the headstock. The X-axis travels perpendicular to the spindle centerline. A negative X (−X) moves the cross slide toward the centerline of the spindle, and a positive X (+X) moves the cross slide away from the spindle centerline. Some machine tool builders mount the cross slide on a slant bed, whereas other manufacturers mount it on a vertical support. Both designs allow the chips to fall free and provide very rigid support.

Turning center movements are controlled through the N/C or CNC unit. Programming is accomplished off-line by an N/C programmer or on the shop floor by the operator through sophisticated CNC interactive graphics systems similar to that shown in figure 7-6. Such shop floor programming controls are geometry-, process-, or motion-oriented and allow the part to be programmed as it is shown on the part print. Foreground/background processing permits programming of a new part while another part is being cut. Other unique shop floor programming features such as part program synchronization, comprehensive menu-selectable tool data, and extensive diagnostics provide a high level of user-friendliness in operator interfaces.

N/C data provides information to the machine tool to operate auxiliary N/C lathe movements such as rotating turrets, circular interpolation, and swing-up tailstocks. Operator-entered CNC input commands can provide the same information to the machine tool as programmed CNC data.

Both absolute and incremental programming are used to position turning center axes. Most older turning centers were limited to incremental positioning only. When incremental positioning is used, the program manuscript form usually contains two extra columns (Z and X) for the programmer to keep track of the absolute dimensions relative to the zero point. Incremental programming today is widely used for continuous loop and other types of repetitive programming. Absolute programming, in contrast, simplifies the effort involved in manual part programming, depending on how the part is dimensioned, and helps ensure accuracy.

Absolute and incremental programming are equally effective. For modern turning centers, the choice is up to the programmer. If the dimensions on the blueprint are given incrementally, the programmer simply programs a G91 for incremental, and the system immediately readies itself to accept incremental input. If workpieces are dimensioned in absolute form, the programmer uses a G90 for absolute input.

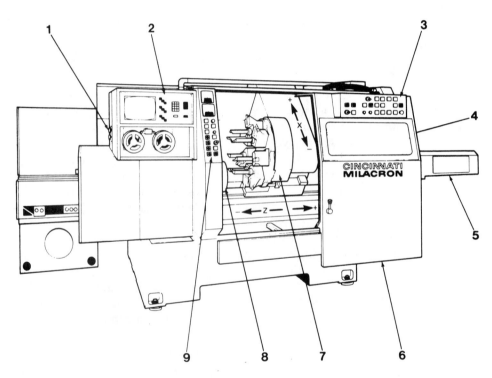

1. HI/LOW CHUCK PRESSURE
2. ACRAMATIC 900TC REMOTE CONSOLE
   AND TAPE READER
3. ROLLING SHIELD PANEL
4. TAILSTOCK QUILL PRESSURE (IF SUPPLIED)
   AND AUTO STEADY REST PRESSURE (IF
   SUPPLIED) ARE MOUNTED ON THE RIGHT SIDE
   OF BED ROLLING SHIELD IN ILLUSTRATION
   HIDES THESE CONTROLS.
5. CHIP CONVEYOR
6. ROLLING SHIELD
7. TURRET
8. CHUCK
9. HEADSTOCK PANEL

FIGURE 7-5
Typical CNC turning center, with axis directions and major components indicated.

FIGURE 7-6
CNC interactive graphics control system. (Courtesy of Cincinnati Milacron Inc.)

OD AND ID OPERATIONS

Regardless of the type of N/C turning center used, a variety of OD and ID operations are performed. In this discussion of OD and ID operations, we will refer to a slant-bed machine with both OD and ID tooling mounted on the same turret-indexing mechanism. Figure 7-5 shows a seven-position turret-indexing mechanism that possesses the capacity for seven OD tools and seven ID tools. Most OD and ID tools have clearance *offsets* to avoid interference with the chuck. In addition, there is automatic compensation of these offsets when tools are changed and when there is a switch from an OD to an ID operation or from an ID to an OD operation.

Qualified toolholders must be used to perform OD operations. The location of a tool insert is held to close tolerances with respect to the rear and opposite sides of the toolholder. This allows holders to be changed without setting gages (see figure 7-7). OD tools are qualified in terms of a standard nose radius for each insert (see figure 7-8). When an insert with the standard radius is used, the intersection of lines parallel to the X- and Z-axes and tangent to the nose radius is located in the same position for all OD tools when the tools are indexed to the machining position. This point is the *common tool point*, shown in figure 7-7. It serves as a common reference point for programming axis coordinates.

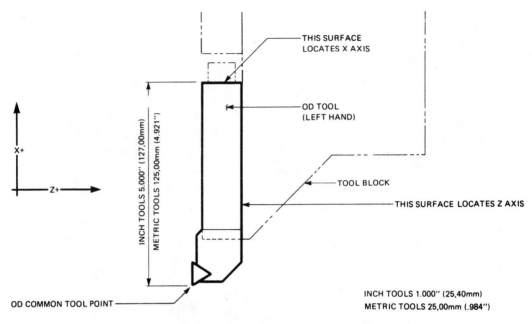

FIGURE 7-7
Qualified OD Toolholder.

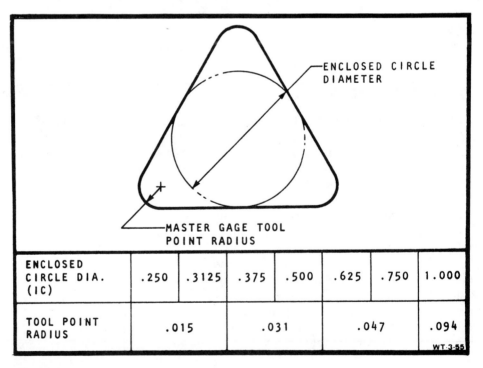

ENCLOSED CIRCLE DIA. (IC)	.250	.3125	.375	.500	.625	.750	1.000
TOOL POINT RADIUS	.015		.031		.047		.094

WT-3-55

FIGURE 7-8
Standard qualified tool point insert radii.

The centerline of ID tools is located on the face of the turret at the same place for inch and metric tool blocks (see figure 7-9).The centerline of ID tools is located at a fixed distance from the OD common tool point along the X-axis. The distance from the ID tool centerline to the tip of the tool (distance A in figure 7-9) varies with the size and construction of the particular tool. For this reason, the distance between the OD common tool point and the ID tool tip of each ID tool must be determined and compensated for when program coordinates for the X-axis are being established. This is illustrated in figure 7-9.

The distance from the tip of the ID tool to the OD common tool point along the Z-axis (distance B in figure 7-9) also varies with the size and adjustment of each tool. These distances must be determined and compensated for when program coordinates for the Z-axis are being determined.

On some N/C turning centers, the distance by which each ID tool is offset from the OD common tool point must be established before the program is written. This must be done to determine axis coordinates and axis movements necessary to avoid interference between the tooling and workpiece. This information must be provided to the operator so that the tools can be set correctly.

To change tools on an N/C turning center, one must program the new turret station along with the tool change code (M06). The turret station and tool offsets are programmed on some CNC turning centers as a four-digit number preceded by the letter T. The first and second digits normally designate the turret station; the third and fourth digits designate the tool offset.

Format: Txxxx

Tool offsets are then dialed in at the control panel. They are used to compensate for tool wear or for minor setup adjustments.

A variety of turning centers offer a four-axis capability in order to perform simultaneous OD and ID cutting operations. Four-axis lathes with individual programmable slides allow for considerable savings because more than one tool is in the cut at a time. This feature is illustrated in figure 7-10. The use of four-axis lathes constitutes a considerable improvement in productivity over conventional turning centers (which have one tool in the cut at a time). However, extreme care must be exercised in programming machines of this technical complexity because the chance for errors and accidents is much higher as a result of the two independent slide movements.

Other turning centers have two spindles with two independent slide motions for the respective OD and ID operations (see figure 7-11). Machines of this nature are also capable of achieving high productivity levels with considerable savings.

Some turning centers (see figure 7-12) are equipped with a robot arm for automatic part loading and unloading. Cutoff bar stock is brought to the machine and completed workpieces taken away by means of a conveyor system synchronized with the robot arm and machine.

As illustrated in figure 7-13, robots are increasingly being used to change workpieces for high production applications. Production requirements today involve a combination of mass production and small lot requirements, resulting in smaller lot production runs that repeat themselves over time.

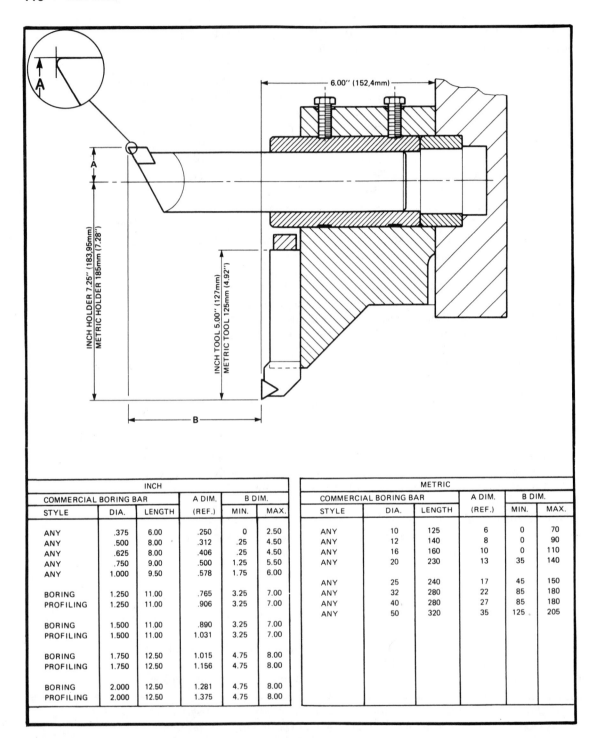

INCH HOLDER 7.25'' (183,95mm)
METRIC HOLDER 185mm (7.28'')

INCH TOOL 5.00'' (127mm)
METRIC TOOL 125mm (4.92'')

6.00'' (152,4mm)

	INCH				
COMMERCIAL BORING BAR			A DIM.	B DIM.	
STYLE	DIA.	LENGTH	(REF.)	MIN.	MAX.
ANY	.375	6.00	.250	0	2.50
ANY	.500	8.00	.312	.25	4.50
ANY	.625	8.00	.406	.25	4.50
ANY	.750	9.00	.500	1.25	5.50
ANY	1.000	9.50	.578	1.75	6.00
BORING	1.250	11.00	.765	3.25	7.00
PROFILING	1.250	11.00	.906	3.25	7.00
BORING	1.500	11.00	.890	3.25	7.00
PROFILING	1.500	11.00	1.031	3.25	7.00
BORING	1.750	12.50	1.015	4.75	8.00
PROFILING	1.750	12.50	1.156	4.75	8.00
BORING	2.000	12.50	1.281	4.75	8.00
PROFILING	2.000	12.50	1.375	4.75	8.00

	METRIC				
COMMERCIAL BORING BAR			A DIM.	B DIM.	
STYLE	DIA.	LENGTH	(REF.)	MIN.	MAX.
ANY	10	125	6	0	70
ANY	12	140	8	0	90
ANY	16	160	10	0	110
ANY	20	230	13	35	140
ANY	25	240	17	45	150
ANY	32	280	22	85	180
ANY	40	280	27	85	180
ANY	50	320	35	125	205

FIGURE 7-9
Relationship between OD and ID tooling.

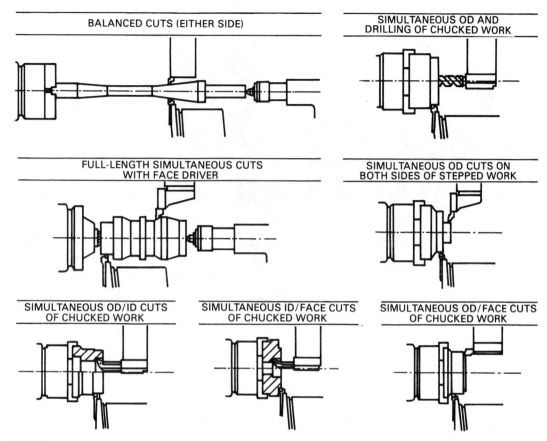

FIGURE 7-10

Four-axis N/C lathe operations, demonstrating simultaneous operations performed on a single-spindle machine.

It is extremely important to note that on any N/C lathe, the turret must be positioned to a location free from interference with the chuck, workpiece, and machine elements before any tool changes are programmed. Failure to comply with this cardinal rule of N/C programming may result in bodily injury and/or machine and tool damage.

FEED RATES

The traverse rate of N/C lathe axes may be programmed in several ways. Usually the options include *vector rapid traverse, feed per minute,* and *feed per revolution.* When the programmed movement requires the traversing of both axes, the axes move simultaneously along a vector path. The rate of travel of each axis is set automatically by the control so that the rate along the vector is equal to the programmed feed or rapid rate, as depicted in figure 7-14.

When a vector rapid traverse is active, the axes move simultaneously from the current position to the command position along a straight vector. Usually,

FIGURE 7-11
Dual-spindle N/C turning center with independent slide movements. (Courtesy of Turning Machine Division, The Warner & Swasey Co., Subsidiary of Bendix Corporation)

FIGURE 7-12
Turning center equipped with a robot arm and conveyer system for automatic part loading and unloading. (Courtesy of Cincinnati Milacron Inc.)

FIGURE 7-13
Floor-mounted robot loading/unloading turning center. (Courtesy of Cincinnati Milacron Inc.)

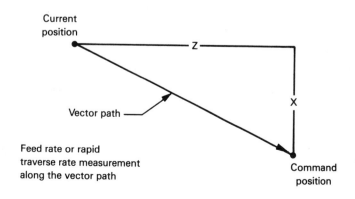

FIGURE 7-14
Vector movement.

the traverse rate along the vector is around 400 IPM or higher. The rapid traverse rates may be modified by the operator, usually by means of a *feed rate override* percent switch.

When the required preparatory function for feed-per-minute feed rate is programmed (G94), the feed rate is usually independent of the spindle speed. The axis feed rate for most turning centers is controlled by a four-digit number preceded by the letter F. The maximum and minimum feed rates will vary from manufacturer to manufacturer.

Although decimal point programming is the predominant programming format in use today, many turning centers are still in use that utilize the interchangeable variable block format. The format for programming the feed rate in the interchangeable block format would be similar to the following:

Inch format: Fxxx.x

Metric format: Fxxxx.

If the programmed feed rate exceeds the allowable feed rate range, the cycle will continue, but the feed rate will be set to the maximum allowable rate. If the feed rate is less than the minimum allowable rate, it will be set to the minimum allowable value.

Another preparatory function (G95) may be used to specify feed rate in terms of vector feed per revolution of the spindle. In this mode, the rate of axis travel varies as a function of spindle speed. The feed rate is programmed by an F word, usually in increments of .0001 in. per revolution of the spindle.

Inch format: F.xxxx

Metric format: Fx.xxx

If a programmed feed-per-revolution rate drops out of its allowable range, the rate will be set to either the maximum or minimum allowable value.

SPINDLE SPEEDS

Spindle speeds are normally programmed on N/C turning centers with a four-digit number preceded by the letter S. The spindle speed can be programmed in either direct RPM coding or in the constant surface speed (CSS) mode.

Format: Sxxxx. (RPM)

Sxxxx. (feet per minute)

Sxxxx. (meters per minute)

Maximum and minimum speeds available will vary, depending on the size and type of N/C lathe.

Other types of N/C lathes use only a two- or three-number code. The code usually refers to a table of speeds for a particular lathe. In addition, different speed ranges may be used, thereby increasing the number of available spindle speeds. Most modern CNC systems maintain a built-in system check to make sure the active spindle S word is within the allowable headstock range. A spindle speed word is also normally programmed in every block containing a headstock range change. Some N/C turning centers will display an error message and stop the cycle if the programmed spindle speed is not within the designated headstock range.

Most N/C lathes use some type of CSS or feet per minute (FPM) feature, as discussed in Chapter 5. CSS automatically varies the spindle rotation rate as a function of the X-axis position to maintain the programmed value of workpiece surface speed at the point of the tool. CSS is normally input by coding a G96. Switching to direct RPM can be accomplished by programming a G97.

FORMAT INFORMATION

Most modern CNC turning centers accept the decimal point programming format. Other CNCs accept the word address, or interchangeable, format with either the EIA (BCD) or ASCII coding.

The following list explains words used for a typical CNC turning center. Some words are standard for any CNC turning center, whereas others may be EIA or manufacturer-specific. However, not all of these words are used for every turning center CNC and its programs.

N   The sequence number is composed of up to four digits preceded by the letter N (Nxxxx). This word is used to indicate which block of information is being processed by the control.

G   The preparatory function code is a two-digit number preceded by the letter G (Gxx). These codes are used throughout the program to define the various modes of operation.

X/Z   Axis dimensions are used to denote the position of the axes. The axis values are addressed in decimal point form and are preceded either by the letter X or Z (X/Z±xxx.xxxx). The sign denotes either the direction of travel (for the incremental mode), or the position relative to the program zero (for the absolute mode).

I/K   Center point coordinates are used to define the center location when circular arcs are being programmed. Center point coordinates are addressed in decimal point form preceded by a plus (+) or minus (−) sign and the letter I or K (I/K±xxx.xxxx). The center point coordinates can be either absolute or incremental, depending on the input mode. I represents the X-axis, and K represents the Z-axis.

I/K   The axis feed rate for threading is controlled by programming of a lead value. This is normally an unsigned decimal number preceded by the letter I or K (I/Kxx.xxxxx). Values for thread lead are not affected by whether an absolute or incremental input mode is in effect. The lead values are always positive, and the sign is not programmed. Programming a negative value will usually result in a program error. I represents X-axis lead, and K represents Z-axis lead.

A   The rapid traverse increment is programmed with an unsigned number preceded by the letter A (Axxx.xxxx). This word is used with the automatic repeat cycle feature to define the incremental rapid approach of the tool to the work.

F   The axis feed rate is controlled by a decimal point number preceded by the letter F (Fxx.x for IPM or F.xxxx for IPR). Feed rates may be programmed in either distance of travel per minute or distance per revolution of the spindle, depending on the selected preparatory function.

R   The radius dimension used for CSS programming is a decimal point number preceded by a plus (+) or minus (−) sign and the letter R (R±xxx.xxxx). The R dimension is always an incremental value measured from the spindle centerline to the tool tip.

V   The tool retract feature is programmed with a two-digit V word (Vxx). This feature programs a tool retraction along an interference-free path. The two digits of the V word represent the X- and Z-axes, respectively. The value of the digit that is programmed determines the direction and distance the tool will travel when the operator initiates the tool retract feature.

S   Spindle speeds are programmed with a four-digit number preceded by the letter S (Sxxxx.). The spindle speed can be programmed in either the direct RPM mode or in the CSS mode.

T   The turret station and offsets are programmed with a four-digit number preceded by the letter T (Txxxx). The first and second digits usually identify the turret station, and the third and fourth digits represent the offset.

C   The C word is used to define the total number of thread starts and the thread start to be machined for machining of multiple-start threads.

M   The miscellaneous function codes are two-digit codes preceded by the letter M (Mxx). These codes are used throughout the program to perform such functions as spindle starting and stopping, coolant control, and transmission range selection.

OPERATIONS PERFORMED BY CNC TURNING CENTERS

The absolute input mode is selected by programming of a G90 word. In the absolute mode, all dimensions input into the control are referenced from a single zero point. The algebraic signs (+ and −) of absolute dimensions denote the position of the axis relative to the zero point. They do not directly specify the direction of axis travel. Some N/C units assume the G91 incremental mode when starting or when data reset operations are performed. When the program is written using the absolute mode, the G90 code should be programmed at the beginning of every operation that uses a new tool.

The incremental input mode is selected by programming a G91 word. In this mode, all dimensions input into the control are referenced from the present axis position. The input dimensions denote the distance to be moved, and the sign

(+ or −) specifies the direction of axis travel. If an entire program is written in the incremental mode, the X- and Z-axes must be returned to the program start point upon completion of the program. If this is not done, the axes will not be in the correct position for the start of the next workpiece. This incorrect placement can cause interference between the tool and workpiece or other components, resulting in tool breakage, damage to the machine, and personal injury.

Linear Interpolation

The G01 linear interpolation preparatory function commands the slides to move the tool in a straight line from the current position to the command position. The rate of traverse is measured along the vector connecting the two points and is equal to the programmed feed rate. This mode of operation is used for turning, drilling, or boring straight diameters; facing shoulders; and turning or boring chamfers and tapers.

The programmer generates the tool path by programming the coordinates of the imaginary tool point of the tool insert radius (see figure 7-15). When straight diameters are being turned or bored or when facing cuts are being made, the programmed coordinates represent tool tangent points created by constructing lines parallel to the machine axes.

When a qualified OD tool with the proper tool insert radius is being used, the imaginary tool point and the common tool point coincide. However, when chamfers or tapers are being turned (see figure 7-16), the tool tangent point is not one of the points shown in figure 7-15. Therefore, in such situations the machine axes must be offset to compensate for the new tangent point location. The amount of compensation depends on the angle of the taper and the radius of the tool insert.

Figure 7-16 illustrates the position of the imaginary tool point at the start and end of the chamfer. It also shows the path that the imaginary tool point follows during the cut. To maintain workpiece tolerances, the axes must be offset at both the start and end of the chamfer by amounts equal to A and B, respectively. When the angle of the chamfer is 45°, A and B are equal.

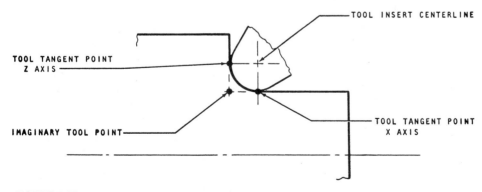

FIGURE 7-15
Tool radius tangent points (cuts parallel to a machine axis).

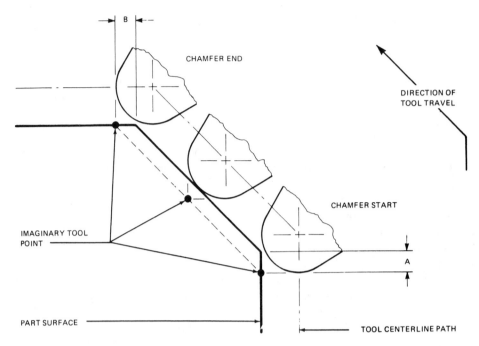

FIGURE 7-16
Imaginary tool point at start and end of a chamfer.

Circular Interpolation

Circular interpolation on an N/C turning center involves moving the tool in a circular arc along a path generated by the control system. The rate of travel is constant around the arc, with a *tangential vector feed rate* equal to the programmed feed rate. Circular interpolation is specified by a G02 preparatory function for the clockwise direction and by G03 for the counterclockwise direction. Coordinate information is also programmed to define the start point, the end point, and the center point (I is the X coordinate value, and K is the Z coordinate value).

Figure 7-17 illustrates motion of the tool from the start point through a circular arc to the end point. Since the tool travels in a clockwise direction, G02 preparatory function is used. At the start point, the centerline of the tool nose is on the arc centerline in the Z-axis. At the end point, the tool nose centerline is on the arc centerline in the X-axis. If the absolute mode is in effect, the I and K programmed center point coordinates are referenced from the program zero. They are offset from the part radius center point by an amount equal to the tool nose radius (TNR). The information required to position the tool through the circular arc movement in the interchangeable variable block format, is

N420   G02   X23188   Z20000   I23188   K29688

Figure 7-18 illustrates motion of a tool from the start point through a circular arc to the end point. Since the tool travels in a counterclockwise direction, a

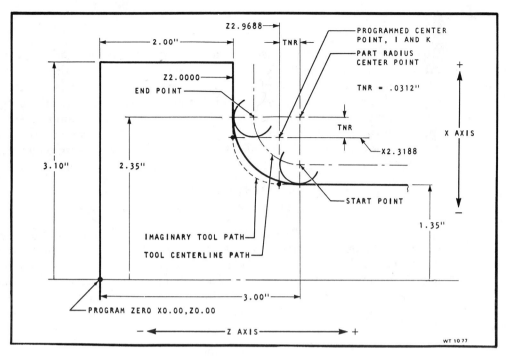

FIGURE 7-17
Inside arc (absolute mode); not drawn to scale.

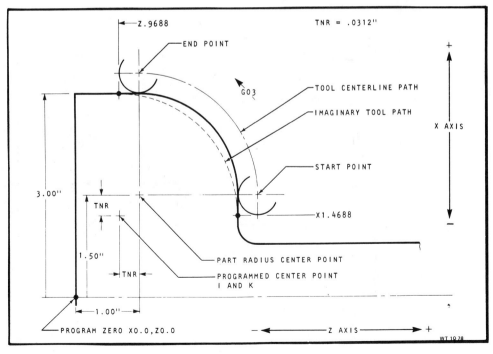

FIGURE 7-18
Outside arc (absolute mode); not drawn to scale.

G03 preparatory function is used. At the start point, the tool nose centerline is on the arc centerline in the X-axis. At the end point, the centerline of the tool nose is on the arc centerline of the Z-axis. If the absolute mode is in effect, the I and K programmed center point coordinates are referenced from the program zero. They are offset from the part radius center point by an amount equal to the tool nose radius. The information required to position the tool to the start point is shown below in block N430. Block N440 shows the interchangeable variable block format information required for the circular arc movements.

N430   G01   X14688   Fxxx

N440   G03   X30000   Z9688   I14688   K9688

In figure 7-19, the tool is moving through a circular arc in the incremental mode. A G02 preparatory function is used, since the tool moves in a clockwise direction. For the incremental mode, the X-axis and Z-axis incremental departure values (Xd and Zd) for 90° inside arcs that start on an axis crossover point are calculated as follows. (In circular interpolation, the crossover point is defined as the point where the end of one 90° quadrant of a 360° circle becomes the starting point for the next 90° quadrant.)

$$Xd = \text{Part radius} - \text{TNR}$$
$$Zd = \text{Part radius} - \text{TNR}$$

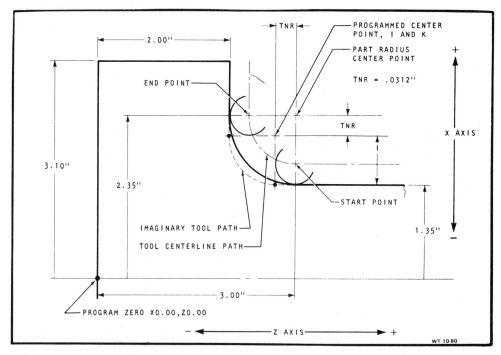

FIGURE 7-19
Inside arc (incremental mode); not drawn to scale.

The I and K dimensions are incremental values measured from the imaginary tool point to the arc center point when the tool is positioned at the start point. In this example, the I value is determined as follows:

$$I = \text{Part radius} - \text{TNR}$$
$$= 1.0000 - 0.0312$$
$$= 0.9688 \text{ in.}$$

The K value is equal to zero because the tool nose radius is on the part radius centerline in the Z-axis.

The example in figure 7-20 illustrates motion of a tool through an outside circular arc in incremental mode. Since the tool moves in a counterclockwise direction, a G03 preparatory function is used. The X-axis and Z-axis incremental departure values for 90° outside arcs, that start on an axis crossover point are calculated as follows:

$$Xd = \text{Part radius} + \text{TNR}$$
$$Zd = \text{Part radius} + \text{TNR}$$

The I and K dimensions are incremental values measured from the imaginary tool point to the arc center point when the tool is positioned at the start point. In this example, the K value is

$$K = \text{Part radius} + \text{TNR}$$

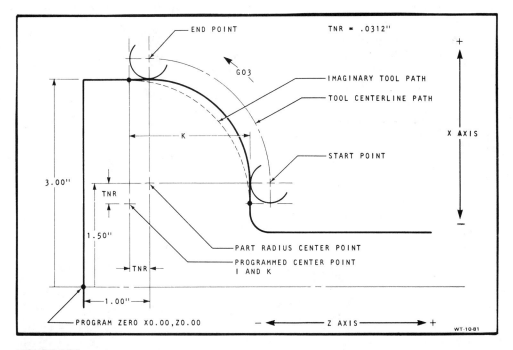

FIGURE 7-20
Outside arc (incremental mode); not drawn to scale.

The I value is equal to zero because the tool nose radius centerline falls on the part radius centerline in the X-axis.

Threading

Modern turning centers are capable of machining constant-lead straight, tapered, and multiple-start threads. The tool is first positioned to depth and to the correct starting distance away from the workpiece. A G32 or G33 block is then programmed to cut the thread. The tool is retracted and returned for the next pass. The process is repeated, making successively deeper cuts until depth is reached. Each of these movements normally requires a separate block of information.

Lead is defined as the amount a thread advances in one revolution of the spindle. The slide feed rate is controlled in the constant-lead threading mode by I and K words in the program. In this case, I designates lead in terms of threads per inch, along the X-axis, and K designates lead along the Z-axis. I and K words must be programmed in every block that contains a threading command.

The expression for lead is as follows:

$$\text{Lead} = \frac{1}{\text{Threads per inch}}$$

Lead and the number of threads per inch are always considered to be positive. (Note that lead can also be defined in terms of threads per millimeter for metric units.)

Before any threading operation begins (that is, before the tool point enters the thread), the tool point must be positioned away from the workpiece (see figure 7-21). The minimum starting distance is

$$\text{Starting distance} = (\text{RPM} \times \text{Lead} \times 0.006) + \text{CO}$$

where CO is the compound in-feed offset (the offset generated by advancing the tool at a 29° angle). The compound in-feed offset is calculated as

$$\text{CO} = \tan 29° - (\text{Full thread depth} - \text{Depth of first pass})$$

A calculation similar to that for the starting distance must be performed for each threading pass when compound in-feed is used to advance the tool to depth. Figure 7-22 illustrates the compensation required for the Z-axis when an 0.0080-in. depth-of-cut pass is made.

When the threads end near a shoulder (see figure 7-23), space must be provided for the slide to stop. The minimum stopping distance is computed as follows:

$$\text{Stopping distance} = \text{RPM} \times \text{Lead} \times .013$$

The following example illustrates the programming of a 1-in.-diameter, eight-threads-per-inch, constant-lead, single thread (see figure 7-24). The program example also uses incremental mode, a compound in-feed of 29° and a 45° pullout.

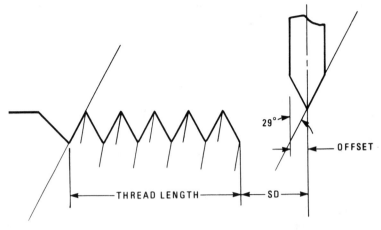

FIGURE 7-21
Starting distance
for compound
in-feed.

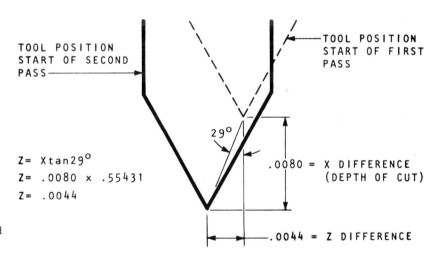

$Z= X \tan 29°$
$Z= .0080 \times .55431$
$Z= .0044$

FIGURE 7-22
Compound in-feed
compensation.

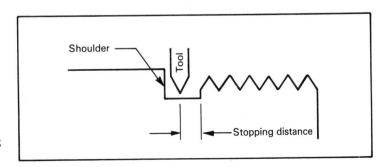

FIGURE 7-23
Minimum stopping
distance.

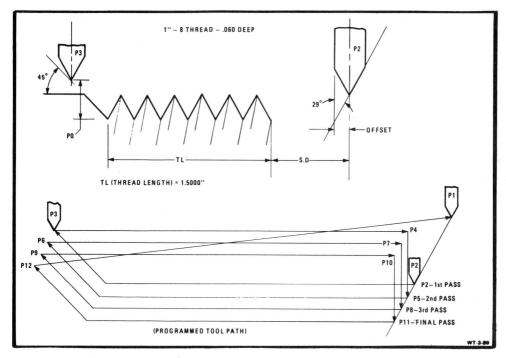

FIGURE 7-24
Example of constant-lead threading, showing the passes required.

Before the program can be written, the following calculations must be made:

$$\text{Lead} = \frac{1}{\text{Threads per inch}}$$

$$= \frac{1}{8}$$

$$= 0.12500 \text{ in. (which is specified as K12500 in the program)}$$

Compound offset

$$\text{CO} = \tan 29° - (\text{Full thread depth} - \text{depth of first pass})$$
$$= \tan 29° - (0.0600 - 0.0300)$$
$$= 0.0166 \text{ in.}$$

Starting distance (with 500-RPM spindle speed)

$$\text{SD} = (\text{RPM} \times \text{Lead} \times 0.006) + \text{CO}$$
$$= (500 \times 0.12500 \times 0.006) + 0.0166$$
$$= 0.375 + 0.0166$$
$$= 0.3916 \text{ in.}$$

X and Z pullout amounts

(In this example, an amount equal to the lead is used for both the X- and Z-axes.)

$$X \text{ pullout} = 0.1250 \text{ in.}$$
$$Z \text{ pullout} = 0.1250 \text{ in.}$$

Z departure distance

$$Zd = TL + PO + (SD - CO)$$

where   TL = thread length
          PO = pullout distance (Z-axis)
          SD = starting distance
          CO = compound offset

$$Zd = 1.5000 + 0.1250 + (0.3916 - 0.0166)$$
$$= 2.0000 \text{ in.}$$

X departure distance

$$Xd = \text{Pullout distance (X axis)}$$
$$= 0.1250 \text{ in.}$$

This sample program is presented to illustrate only the tool path movements in the interchangeable variable block format. Spindle speeds, turret indexes, and other machine-related information have been omitted from the program. The P terms in parentheses are not code; they correspond to locations in the cycle shown on figure 7-24.

```
N480   G90
N490   G00    X10000    Z50000                    (P1)
N500   G91
N510   G00    X-5300    Z-2938                     (P2)
N520   G33    X1250     Z-20000    K12500          (P3)
N530   G00    Z19917                               (P4)
N540   X-1400                                      (P5)
N550   G33    X1250     Z-20000    K12500          (P6)
N560   G00    Z19945                               (P7)
N570   X-1350                                      (P8)
N580   G33    X1250     Z-20000    K12500          (P9)
N590   G00    Z19972                               (P10)
N600   X-1300                                      (P11)
N610   G33    X1250     Z-20000    K12500          (P12)
N620   G00
N630   G90
N640   X10000    Z50000                            (P1)
```

Sample Program Description:

Block N480:

G90 is used to define the starting point P1 in the absolute mode.

Block N490:

G00 is used to rapid the tool to P1.

Block N500:

G91 selects the incremental input mode.

Block N510:

The tool rapids to depth for the first pass. The first cut depth is 0.030 in.

Block N520:

The G33 code selects the threading mode. The X1250 command will cause a 45° pullout at the end of the thread. The Z command includes an additional 0.1250 in. for the pullout. The K word defines the lead.

Block N530:

G00 is programmed to rapid the Z-axis to P4. The Z command includes the compensation value for the 29° compound in-feed. This value is calculated by multiplying tan 29° by the new cut depth.

$$Zd = 2.0000 - \tan 29° \times .0150$$
$$= 2.0000 - 0.0083$$
$$= 1.9917 \text{ in.}$$

Block N540:

The X command rapids the tool to P5. This dimension is obtained by adding the amount of retraction to the depth of the cut for the second pass. The sign is negative because the movement is toward the centerline of the spindle.

$$Xd = 0.1250 + 0.0105 + 0.1400$$

Block N550:

G33 selects the threading mode for the second pass, which ends at P6.

Block N560:

G00 is used to rapid the Z-axis to P7.

Block N570:

The X command rapids the tool to P8. The depth of cut for this pass is 0.010 in.

Block N580:

G33 is used to make the third threading pass, which ends at P9.

Block N590:

G00 is used to rapid the Z-axis to P10.

Block N600:

The tool rapids to P11. The depth of cut for this pass is 0.005 in.

Block N610:

G33 is used to make the final threading pass, which ends at P12.

Block N620:

G00 is programmed to cancel the G33 and to select rapid traverse.

Block N630:

G90 selects the absolute input mode.

Block N640:

The X and Z commands return the tool to P1.

Some N/C turning centers provide a finish threading feature that allows for programming of rough and finish threading passes that use different spindle speeds without introducing thread form errors. The threads may be rough machined at a low spindle speed and finished at a high speed. This is accomplished by compensating for changes in what is called *following error.* When a slide moves, the actual motion lags behind the command signal. This lag—the following error—is characteristic of all servo systems and is a function of feed rate. The feed rate during threading is determined by spindle speed (assuming a given lead). Changes in spindle speed will result in slight shifts of the tool position as a result of following error changes. This in turn produces an error in thread form.

However, when the finish threading feature is used, the control measures the following error of the initial pass. When the finish pass is made, the control automatically adjusts the slide position to maintain the same following error as during the original pass. The thread form error is thereby eliminated.

SYSTEM SUBROUTINES

A *subroutine* is a set of commands or instructions that are identified and stored in the CNC system. When called upon, these instructions are put into action. The process of activating a subroutine—sometimes referred to as a *macro*, or a program within a program—entails calling for this set of blocks or instructions.

Most CNC units contain stored parametric variable subroutines as an optional feature. These subroutines contain frequently used instructions that are stored in the control but external to the program. Once stored, the subroutine is viewed by the control in the same manner as a part program. Usually a maximum number of subroutines, part programs, or any combination of the two may be stored, depending on control specifications. The number of blocks that can be programmed in a subroutine is limited only by the total storage capacity of the manufacturer's control.

Subroutines are made active by a call statement in the part program. The subroutine can be repeated many times with a single call statement and can be called any number of times by the part program.

A parametric subroutine may be programmed with variable commands so that it may be used for a variety of workpiece configurations. Ten or more variable commands can usually be used in a subroutine program. The variable commands are assigned values by the call statement of the part program.

MCUs containing parametric subroutines are usually stored separate from, but in the same area as, a part program. Subroutines may be permanently or temporarily stored in the CNC's memory. A permanently stored subroutine is loaded into memory by means of a multiple-program store feature. A temporarily stored subroutine is normally loaded at the beginning of the part program.

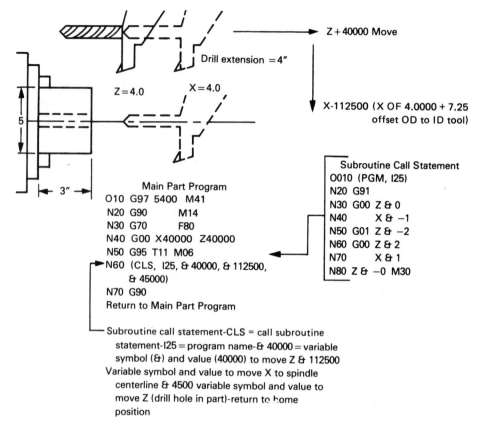

FIGURE 7-25

Example of a permanently stored subroutine and the actions it specifies.

A typical example of a permanently stored subroutine is given in figure 7-25. This subroutine is used to offset in Z, move into position in X, drill a hole in the part, and return to the home position (X = 4.0 in. and Z = 4.0 in.). The subroutine defines the tool movements required to drill the part. The subroutine call statement is programmed to repeat the subroutine as many times as specified by the main program.

WORK HOLDING AND WORK CHANGING

Work-holding and work-changing equipment for turning centers have been improved and automated to further reduce machine down time for part loading and unloading. This equipment generally consists of automatic chuck and chuck-jaw changing systems, countercentrifugal chucks, and automatic pallet-changing systems. It adds to the machine productivity and utilization of stand-alone turning centers as well as those used in flexible manufacturing cells and systems.

FIGURE 7-26

Countercentrifugal chucks incorporate internal counterweights that pivot to offset centrifugal force when the chuck rotates, thereby increasing the grip on the workpiece. (Courtesy of ITW Woodworth)

Traditional work-holding equipment (chucks and special-application face-plate drivers) must withstand the tremendous cutting forces generated by the metal-removal process and must also compensate for the centrifugal forces that, at high rotating speeds, counteract the clamping power provided by the work-holding equipment (such forces create a tendency for the chuck to open).

Countercentrifugal chucks (see figure 7-26) practically eliminate the constant and dangerous centrifugal force, which tends to open the chuck jaws during rotation, by incorporating the use of internal counterweights of wedge, lever, or wedge/lever actuation types. These counterweights pivot as the chuck rotates so that centrifugal force tends to increase the grip on the part as the chuck rotates faster, thereby offsetting the outward-developed centrifugal jaw forces.

In order to further increase a turning center's up time, chucks and chuck jaws need to be changed quickly. Traditionally, changing chuck jaws could take 20 to 30 minutes, but new automated quick-change designs allow jaws, chucks, or pallets to be changed in less than two minutes.

Modern jaw-changing systems (see figure 7-27) are controlled hydraulically, but they generally have an electronic stroke control to select, set, and secure the desired clamping range and position. With this type of system, a 12-in. diameter chuck, utilizing five sets of quick-change jaws, has a flexible clamping range from approximately 1 to 9 in. in diameter. In this type of environment, automatic changing of work-holding tooling considerably reduces human intervention for changeover and provides increased flexibility for short-run production.

In some applications, workpiece variety may not permit an automated jaw change because too great a size variation exists between different workpieces. In such cases, the chuck itself may be changed, as shown in figure 7-28. This can be accomplished by a hydraulic locking/releasing mechanism and automated crane or through a robotic application.

Pallet-changing systems are sometimes utilized for the machining of complex rotational parts. These systems typically require less than one minute for changeover and might use a gantry robot, as shown in figure 7-29. The robot

FIGURE 7-27
Modern jaw-changing system.
(Courtesy of Mazak Corp.)

rotates the pallet from the horizontal to the vertical plane and positions the loaded pallet precisely on the turning center's spindle. Hydraulic clamp fingers are actuated to securely lock the pallet to the spindle face. Pallet-changing systems expand flexibility by allowing the entire pallet and part assembly to be moved to other machines for drilling, milling, or tapping operations.

FIGURE 7-28
Chuck-changing systems are used in applications where workpiece size variation is too great for automated jaw changing. (Courtesy of ITW Woodworth)

FIGURE 7-29
Complete pallet-changing system with a gantry robot. (Courtesy of Cincinnati Milacron Inc.)

AUTOMATED FEATURES AND CAPABILITIES

By providing automated features and capabilities, turning centers are now capable of performing unattended operations and functions that previously were performed manually by skilled operators. These functions are discussed in the following subsections.

Probing

A probe, sometimes called a touch-trigger probe (see figure 7-30), is a very effective measuring device with applications to a wide variety of situations. The probe is a high-precision switch. The probe stylus is attached to the switch, and when the stylus is deflected (see figure 7-31), the switch contacts close, allowing current to pass. The machine control must be equipped with the proper software, which continually scans the input to the probe; when a contact closure is detected, the probe captures the current values in the machine's active position register. The values from the active position register can be compared with the expected values in the part program to calculate a dimensional deviation. This deviation can then be entered as an offset to correct or regrid the machine axes. Probing accuracy depends on the repeatability of the switch, the accuracy of the

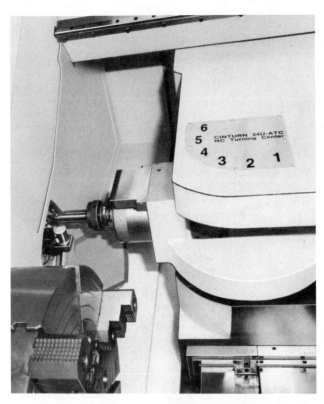

FIGURE 7-30
Touch-trigger probing is a widely used example of postprocess, on-machine gaging. (Courtesy of Cincinnati Milacron Inc.)

FIGURE 7-31
The probe is a high-precision switch whose accuracy depends on the repeatability of the switch, stored machine register values, and control register accessing speed. (Courtesy of Cincinnati Milacron Inc.)

values in the machine position registers, and the speed at which the control can access the values in the registers. Accuracy can be increased by touching the probe to the reference surface, usually the machine's chuck, for recalibration.

Live Spindle Tooling

Rotary tooling (see figure 7-32) is one of the most important new features associated with advanced turning centers. Available on both horizontal and vertical machines, it permits milling, drilling, and tapping on the part face and outside diameter while the workpiece remains stationary and clamped in the chuck or on a face plate. Operations that can be performed with rotary tools can be divided into two broad categories: (1) those requiring only a spindle index to an oriented stop, and (2) those in which spindle rotation (C-axis) is a feed motion moving simultaneously with X, Z, or both linear axes. Separate drilling and milling machines for secondary operations on turned workpieces, along with the additional piece-handling time, can be eliminated in many cases. In addition, accuracy is improved because milling and drilling are done in the same setup, eliminating refixturing and its effects on concentricity. However, the addition of milling, drilling, reaming, and tapping cycles allows the completion of workpieces on the turning center at a cost of increased cycle and programming times.

Tool-Changing Systems

The typical N/C turning center's 12-to-16-tool turret cannot provide enough tools and tool inserts for flexible, unattended turning operations. Tool changers on N/C turning centers have been available since the early 1970s, with completely automated tool handling becoming a reality in the 1980s. Tool-changing systems vary from those that change only the cutting tool heads to those that change the entire tool holder (see figure 7-33). In some cases, the tool assemblies are similar to and interchangeable with machining center tool assemblies and can reside in an attached tool storage drum or magazine (see figure 7-34). Many of these tool-changing and tool storage turning centers will swap tools to the ram either by moving the ram laterally to the tool drum or magazine or by removing the tool and transporting it to the ram. The principal benefits of automated turning center tool-changing systems are

- Reduction in setup time
- Reduction in insert-changing time
- Added machine flexibility
- Improved accuracy and a reduction in errors through reduced manual intervention
- Increased machine utilization

Tool Monitoring and Sensing

Tool wear and breakage (see figure 7-35) require operator attention as often as every three minutes in turning operations. Tool-monitoring and tool-sensing systems substitute for the skilled operator's eyes and ears and signal the need

FIGURE 7-32
Turning center equipped
with rotary (live spindle)
tooling. (Courtesy of
Cincinnati Milacron)

FIGURE 7-33
Automated turning center tool-changing systems
are available, ranging from those that change only
the cutting tools to those that change the entire tool
holder. (Courtesy of Mazak Corp.)

FIGURE 7-34
On some automated turning centers, tool capacity is increased through the addition of a tool storage chain or magazine. (Courtesy of Cincinnati Milacron Inc.)

FIGURE 7-35
Tool monitoring and sensing help to considerably reduce operator attention to cutting tool wear and breakage conditions. (Courtesy of Kennametal Inc.)

to replace worn or damaged tools. The simplest form of tool monitoring is not a true tool-monitoring system. It simply records the actual cut time of each tool, compares that with the preprogrammed limits for the tool's life, and signals for a tool change when the preprogrammed limit has been reached. Other types of tool wear systems include the following:

- *Horsepower sensing:* Based on the fact that a worn tool draws more power than a sharp tool, this type of tool-wear system measures the load on the main spindle drive motor. If the horsepower exceeds the programmed limits, indicating a worn tool, a tool change is signaled.

- *Acoustic-emission sensing:* This type of tool-wear/breakage subsystem operates on the principle that cutting processes emit acoustic pulses at ultrasonic frequencies. When a tool is about to break, acoustic emissions increase up to five times their normal value. When the acoustic-emission sensor detects a rapid increase in acoustic emissions relative to preprogrammed values, the feed can be stopped and a tool change initiated.

- *Learn-mode:* This is a type of feed-force–monitoring system that memorizes the sensor signal values from sharp tools. If a predetermined percentage increase in the feed force is exceeded, a tool change to replace the worn tool at the end of the operation is initiated. In the event of tool breakage, the monitor instantly senses the sudden force increase and signals the control to stop the feed and change to a new tool.

- *Force monitoring:* This type of system typically monitors the force on ball-screw drives along the X- and Z-axes. Sensors that support X- and Z-axis ball-screw drives and bearings measure fluctuations in the feed force and, when these values exceed a predetermined level, initiate a tool change.

EXAMPLE PROGRAMS

The following programs illustrate the machining operations for typical sample parts. The final part for the first example is shown in figure 7-36*a*. The following code, given in the decimal point programming format, would be used for this part (explanations of the code segments are also given).

Refer to figure 7-36*b*.

```
GO TO1O1 M8        Change to tool 1
G50 S3000 M42      Set spindle limit (S3000), high range (M42)
G96 S250 M3        Set constant surface speed (S250), spindle on
GO Z.1 X2.7        Rapid to start point 0.100 in. from part
G1 Z0 F.015        Feed in to Z0
X–.05              Face part 0.050 in. below center
```

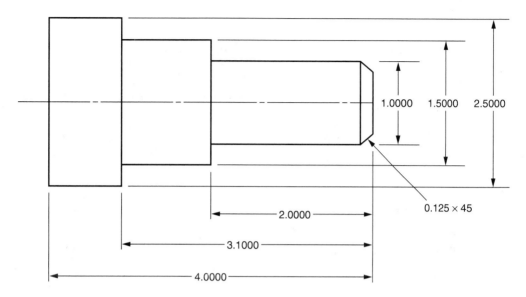

FIGURE 7-36a
Finished part for first example.

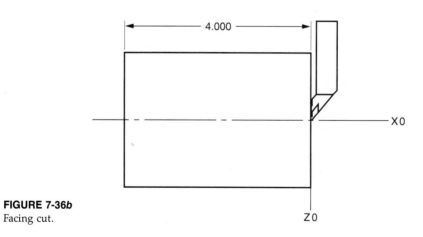

FIGURE 7-36b
Facing cut.

Refer to figure 7-36c.

GO Z.1 X2.7	Rapid to start point
G90 X2.45 Z–3.085 F.010	Activate canned cycle (G90): OD, ID turning; take cuts of
X2.4	0.100 in. off diameter to a Z depth of −3.085 in., leaving
X2.3	extra 0.015 in. on diameter and linear dimensions at
X2.2	last step
X2.1	
X2.0	
X1.9	
X1.8	
X1.7	
X1.6	
X1.515	

Refer to figure 7-36d.

X1.4 Z−1.985	Continue to take cuts of 0.100 in. on diameter but to
X1.3	a Z distance of −1.985 in.
X1.2	
X1.1	
X1.015	

Refer to figure 7-36e.

GO X.75	Rapid to start point 1
G1 ZO F.010	Feed to Z0 at start point 1
X1.OO Z−.125	Feed to point 2
Z−2.000	Feed to point 3
X1.5	Feed to point 4
Z−3.100	Feed to point 5
X2.500	Feed to point 6
Z−4.000	Feed to point 7
X2.7	Feed clear of part
GO X6.000 Z6.000	Rapid to a safe distance
M30	Reset program

The finished part for the second example is shown in figure 7-37. This example demonstrates a rough face, rough turn, and finish profile operation. It does not illustrate all phases of machining operation, programming techniques, or optional equipment. The code for this example (in the word address/interchangeable, or compatible, format) is given below. The bar stock for this example is 4.0 in. × 6.5 in. Tool T01 is used to rough face and rough turn. Tool T04 is used for the finish profile operation (0.0468-in. TNR).

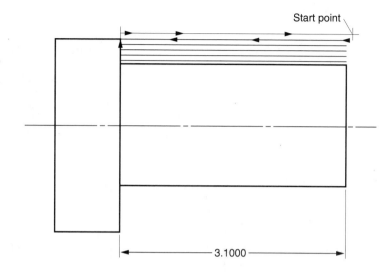

Start point

3.1000

FIGURE 7-36c
Rough turn using
G90 and 0.100-in.
depth of cut.

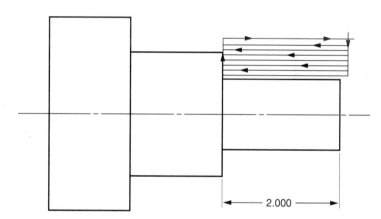

2.000

FIGURE 7-36d
Continuation of
rough turn, 0.100-in.
depth of cut.

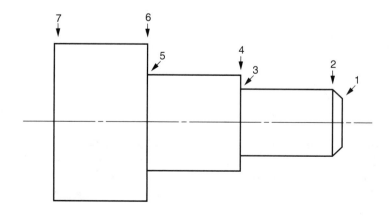

FIGURE 7-36e
Finish turn part.

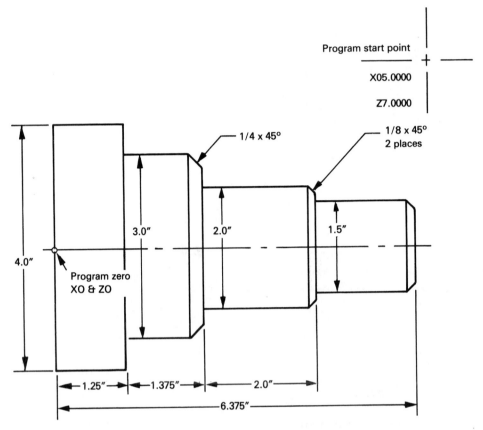

Program start point

X05.0000

Z7.0000

1/4 x 45°

1/8 x 45°
2 places

3.0"

2.0"

1.5"

4.0"

Program zero
XO & ZO

1.25"

1.375"

2.0"

6.375"

FIGURE 7-37
Example part for rough face, rough turn, and finish profile. (Courtesy of Cincinnati Milacron Inc.)

BAR STK. 4.0" x 6.5
T01 (ROUGH FACE AND ROUGH TURN)
T04 FINISH PROFILE (.0468 TNR)

O10	G97	S300	M41	300 rpm — low-gear range
N20	G90		M13	Absolute positioning, spindle on w/coolant
N30	G70			Inch programming
N40	G00	X50000	Z70000	Rapid to program start pt.
N50	G95	T0101	M06	Inches per rev. — index to tool 1 + assn. offset #1
N60	G96	S400	R50000	Constant surface speed (400 sfm)
N70	G00	X22000	Z63850	Rapid to position to rough face

N80	G01	X-312	F150	Face end of stock (leave .010″ for finish)
N90			Z65000	Feed away from end of stock
N100	G00	X17500		Rapid to pos. for rough turn (.25″depth) leave .010 on shldr. for finish
N110	G01		Z13500	Make 1st rough turn to 1.25 shldr.
N120		X18500	Z14500	Feed + .100 in X + Z (clearance away from part)
N130	G00		Z65750	Rapid out in Z-axis .200 from end of part
N140		X16000		Rapid down to next depth of cut
N150	G01		Z13500	Make second rough turn to 1.25 shldr.
N160		X16100	Z14500	Feed + .100 in X + Z (clearance)
N170	G00		Z65750	Rapid out in Z .200 from end of part
N180		X12500		Rapid down to next depth
N190	G01		Z27250	Make rough turn to 2.625 shldr.
N200		X16100	Z22750	Rough 1/4 x 45° chamfer
N210	G00		Z65750	Rapid out .200 from end of part
N220		X11000		Rapid down to next depth
N230	G01		Z27250	Make rough turn to 2.625 shldr.
N240		X12000	Z28250	Feed + .100 in X + Z (clearance)
N250	G00		Z65750	Rapid out .200 from end of part
N260		X7250		Rapid to next depth
N270	G01		Z64750	Feed in to .100 from end of part
N280		X8500	Z63500	Rough 1/8 chamfer on end of part
N290			Z47250	Feed back to 4.625 shldr.
N295		X8850		Feed up 4.625 shldr.
N300		X11000	Z44000	Rough 1/8 x 45 chamfer on 4.625 shldr.
N310		X12000		Feed up .100 to clear part
N320	G00	X50000	Z70000	Rapid turret back to prog. st. pt.
N330	G97	S1000	M42	Direct rpm (1000) — high-gear range
N340	G90			Absolute positioning
N350	G70			Inch programming
N360	G00	X50000	Z70000	Restate current pos. for tool index
N370	G95	T0404	M06	Inches per rev. — index to tool #4 + assn. offset #4
N380	G96	S1000	R50000	Constant surface speed (1000 sfm)
N390	G00	X-470	M13 Z65750	Rapid to ₵ and .200 from end of part
N400	G01	F80	Z63750	Feed in to end of part to begin finish profile
N410		X5976		Feed from ₵ to start of chamfer on end of part
N420		X7500	Z62226	Cuts chamfer on end of part (1/8 x 45°)
N430			Z46250	Feed back in Z to 4.625 shldr.
N440		X8476		Feeds up 4.625 shldr. to start of 2nd 1/8 x 45°C
N450		X10000	Z44726	Cuts 1/8 x 45°C on 4.625 shldr.
N460			Z26250	Feeds across 2.0″ dia. to 2.625 shldr.
N470		X12226		Feeds up 2.625 shldr. to start of 1/4 C
N480		X15000	Z23476	Cuts 1/4 x 45°C up 3.0″ dia.
N490			Z12500	Feeds across 3.0″ dia. to 1.25 shldr.
N500		X21000		Feeds up 1.25″ shldr. + clears str. dia. by .100″
N510	G00	X50000	Z70000	Rapid back to st. pt.
N520		T0100	M06	Cancel out active assignable offset
N530		M30		Ends program (M30 shuts off spindle and coolant — also rewinds program)

The engineering drawing for the third example is shown in figure 7-38*a*. The machining necessary to produce this part is performed in eight operations (shown in figures 7-38*b*, *c*, and *d*). This example is representative of basic turning, boring, and threading operations performed on a CNC turning center using the word address/interchangeable or compatible format. It does not illustrate all phases of machining operation, programming technique, or optional equipment.

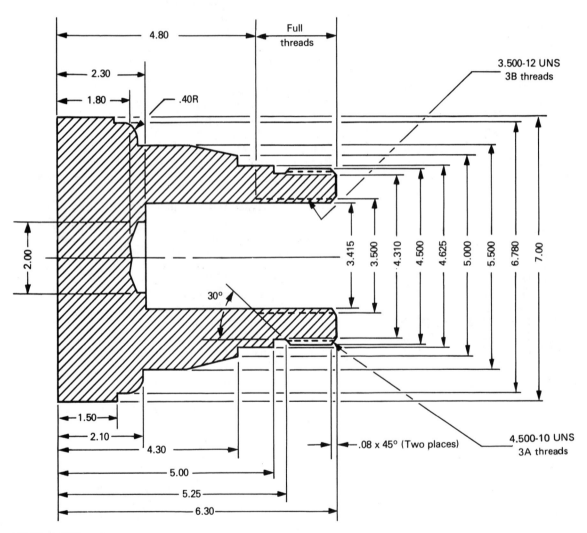

FIGURE 7-38a
Sample part engineering drawing for third example.

First Operation	O10	G90			
	N20	G97 S100 M42			
	N30	G70 M03			
	N40	G00 X50000 Z85000 T0100 M06			
	N50	G95			
	N60	G92 S2500			
	N70	G96 R50000 S600			
	N80	G00 X37000 Z63500 M08			
	N90	G01 X-940 F150			
	N100	Z65500 F600			
	N110	X37000			
	N120	Z63000			
	N130	X-940 F150			
	N140	G00 Z65000			
	N150	X45000			
Second Operation	O160	G90			
	N170	G97 S351 M41			
	N180	G70 M13			
	N190	G00 X45000 Z65000 T0200 M06			
	N200	G95			
	N210	G92 S2500			
	N220	G96 R45000 S600			
	N230	G00 X32600			
	N240	G01 Z19690 F150			
	N250	G03 X34100 Z16062 I28962 K16062			
	N260	G01 X37000			
	N270	G00 Z65000			
	N280	X30100			
	N290	G01 Z21200			
	N300	X32100			
	N310	G00 Z65000			
	N320	X27600			
	N330	G01 Z21100			
	N340	X28962			
	N350	G03 X34000 Z16062 I28962 K16062			
	N360	G01 Z15100			
	N370	X36000			
	N380	G00 X29600 Z65000			
	N390	X25100			
	N400	G01 Z42199			
	N410	X27600 Z32986			
	N420	G00 Z65000			
	N430	X22600			
	N440	G01 Z55989			
	N450	X21650 Z51659			
	N460	Z50100			
	N470	X23225			
	N480	Z43100			

	N490	X26000
	N500	G00 Z130000
Third Operation	O510	G90
	N520	G97 S600 M41
	N530	G70 M14
	N540	G00 X26000 Z130000 T1100 M06
	N550	G95
	N560	G00 X-72500
	N570	G01 Z83000 F150
	N580	G00 Z148000
Fourth Operation	O590	G90 M05
	N600	G97 S1000 M42
	N610	G70 M13
	N620	G00 X-72500 Z148000 T1200 M06
	N630	G95
	N640	G92 S2500
	N650	G96 R10310 S600
	N660	G00 X-71560 Z115000
	N670	G01 Z73100 F150
	N680	X-73750
	N690	G00 Z115000
	N700	X-70310
	N710	G01 Z73100
	N720	X-72500
	N730	G00 Z115000
	N740	X-69060
	N750	G01 Z73100
	N760	X-71250
	N770	G00 Z115000
	N780	X-67810
	N790	G01 Z73100
	N800	X-70000
	N810	G00 Z115000
	N820	X-66560
	N830	G01 Z73100
	N840	X-68750
	N850	G00 Z115000
	N860	X-65835
	N870	G01 Z73100
	N880	X-68023
	N890	G00 Z148000
Fifth Operation	O900	G90
	N910	G97 S846 M42
	N920	G70 M13
	N930	G00 X-68023 Z148000 T1300 M06
	N940	G95
	N950	G92 S2500
	N960	G96 R14787 S800

```
                   N970     G00  X-64752  Z115000
                   N980     G01  X-65735  Z112017   F100
                   N990     Z73000
                   N1000    X-73210
                   N1010    G00  Z148000
Sixth Operation    O1020    G90  M05
                   N1030    G97  S400   M41
                   N1040    G70  M14
                   N1050    G00  X-73210  Z148000   T1400  M06
                   N1060    X-67805  Z103000
                   N1070    G91
                   N1080    G33  X-1500   Z-18668  K8333
                   N1090    G00  Z18613
                   N1100    X1600
                   N1110    G33  X-1500   Z-18668  K8333
                   N1120    G00  Z18629
                   N1130    X1570
                   N1140    G33  X-1500   Z-18668  K8333
                   N1150    G00  Z18640
                   N1160    X1550
                   N1170    G33  X-1500   Z-18668  K8333
                   N1180    G00  Z18646
                   N1190    X1550
                   N1200    G33  X-1500   Z-18668  K8333
                   N1210    G00  Z18654
                   N1220    X1525
                   N1230    G33  X-1500   Z-18668  K8333
                   N1240    G00  Z18670
                   N1250    X1520
                   N1260    G33  X-1500   Z-18668  K8333
                   N1270    G00
                   N1280    G90
                   N1290    Z105000
                   N1300    X45000
Seventh Operation  O1310    G90
                   N1320    G97  S780   M41
                   N1330    G70  M13
                   N1340    G00  X45000   Z105000   T0300  M06
                   N1350    G95
                   N1360    G92  S2500
                   N1370    G96  R45000   S800
                   N1380    G00  X19517   Z65000
                   N1390    G01  X22500   Z62017   F100
                   N1400    Z54373
                   N1410    X21550   Z52728
                   N1420    Z50000
                   N1430    X23125
                   N1440    Z43000
```

```
                     N1450  X24927
                     N1460  X27500   Z33399
                     N1470  Z21000
                     N1480  X29588
                     N1490  G03  X33900  Z16688   I29588  K16688
                     N1500  G01  Z15000
                     N1510  X37000
                     N1520  G00  X45000  Z65000
Eighth Operation     O1530  G90
                     N1540  G97  S300  M41
                     N1550  G70  M14
                     N1560  G00  X45000  Z65000   T0400  M06
                     N1570  X22300  Z66029
                     N1580  G33  Z50300  K10000
                     N1590  G00  X24500
                     N1600  Z65946
                     N1610  X22150
                     N1620  G33  Z50300  K10000
                     N1630  G00  X24500
                     N1640  Z65879
                     N1650  X22030
                     N1660  G33  Z50300  K10000
                     N1670  G00  X24500
                     N1680  Z65835
                     N1690  X21950
                     N1700  G33  Z50300  K10000
                     N1710  G00  X24500
                     N1720  Z65807
                     N1730  X21900
                     N1740  G33  Z50300  K10000
                     N1750  G00  X24500
                     N1760  Z65791
                     N1770  X21870
                     N1780  G33  Z50300  K10000
                     N1790  G00  X24500
                     N1800  Z65780
                     N1810  X21850
                     N1820  G33  Z50300  K10000
                     N1830  G00  X24500
                     N1840  X50000  Z85000
                     N1850  M30
```

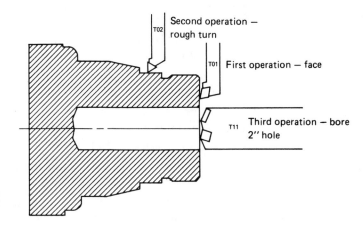

FIGURE 7-38b
First, second, and third operations.

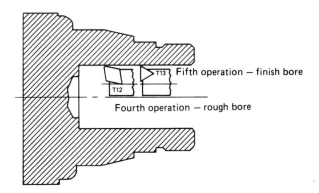

FIGURE 7-38c
Fourth and fifth operations.

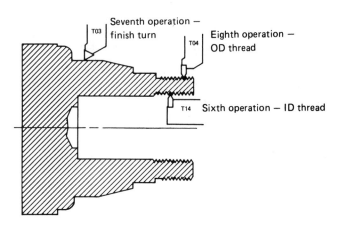

FIGURE 7-38d
Sixth, seventh, and eighth operations.

REVIEW QUESTIONS

1. Explain the relationship between the Z-axis and X-axis on an N/C lathe, including how positive and negative moves are made in each direction.

2. Explain the differences between absolute and incremental programming on an N/C turning center. What is a major factor in determining whether to program in an absolute or incremental mode?

3. How are OD and ID tools accurately located in the tool turret?

4. Discuss the importance of tool offsets. Why are they used?

5. Why is it necessary to know the location of the tip of the tool prior to programming a tool change?

6. Explain the difference between the three types of axis feed rates, and discuss what takes place when a programmed feed rate exceeds an allowable feed rate range.

7. What is meant by constant surface speed (CSS)? Briefly explain how it functions.

8. What turning center operations are performed with linear interpolation moves? How is the tool path information generated?

9. Why is circular interpolation used on N/C turning centers? What are the basic differences between programming an incremental versus absolute circular interpolation move?

10. Why is it important to program Z- and X-axes leads on a threading operation? Are thread lead values affected by absolute or incremental input modes?

11. Why should a space be provided in the part when threading is ended at a shoulder?

12. What is finish threading? How is it accomplished on an N/C turning center?

13. Briefly explain a system subroutine. Are any limitations placed on the number of blocks that are programmed in a subroutine? How?

14. What is a parametric subroutine? What is the difference between a temporary and a permanent stored subroutine? How are each accessed by the N/C part program?

CHAPTER 8

CNC Machining Centers and Programming

OBJECTIVES

After studying this chapter, you will be able to:

- Identify machining center types and list some advantages and disadvantages of each
- Explain primary machining center axes, programming principles, and format information
- Identify the types of operations performed on a CNC machining center
- Discuss pallet, part loading, and programming options
- Describe automated features and capabilities

Known in the 1960s as automatic tool changers (ATCs), machining centers became widely used because of their ability to perform a variety of machining operations on a workpiece by changing their own cutting tools. They began a revolution in tool changing and additional features/capabilities among machine tool builders that continues to escalate today through improvements and enhancements added to the staggering array of machining center options.

Machining centers, even in the early years of "turret drills," began to affect manufacturing operations. In many cases, their adoption served as a shop's introduction to numerical control, and their high productivity frequently forced a reassessment of part setup and processing requirements. Statistically, they have become well accepted as a separate class of machine tool. Today, machining center use continues to expand from stand-alone job-shop applications to flexible manufacturing cells and systems. However, the increased utilization and higher chip-removal rates of automated applications place considerable wear and tear on this relatively new breed of machine tools.

Machining center innovations and developments have brought about the following improvements:

- Improved flexibility and reliability
- Increased feeds, speeds, and overall machine construction and rigidity
- Reduced loading, tool-changing, and other noncutting time
- Greater MCU (machine control unit) capability and compatibility with systems

- Reduced operator involvement
- Improved safety features and less noise

These improvements, driven by increased quality, productivity, and environmental demands along with intense competition among machine tool suppliers, have created part-hungry machine tools able to machine workpieces to exacting tolerances accurately and consistently.

TYPES OF MACHINING CENTERS

Machining centers, just like turning centers, are classified as either vertical or horizontal. Vertical machining centers (see figure 8-1a) continue to be widely accepted and used, primarily for flat parts and situations in which three-axis machining is required on a single part face (such as in mold and die work). Horizontal machining centers (see figure 8-1b) are also widely accepted and used, particularly because parts can be machined on multiple sides in one clamping of the workpiece and because they lend themselves to easy and accessible pallet (movable machine tool tables) shuttle transfer when used in a cell or FMS (flexible manufacturing system) application.

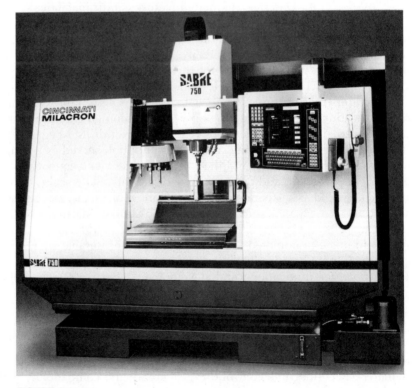

FIGURE 8-1a
Vertical machining center. (Courtesy of Cincinnati Milacron Inc.)

FIGURE 8-1*b*
Typical horizontal machining center. (Courtesy of Cincinnati Milacron Inc.)

Typical motions of machining centers involve the primary axes of X, Y, and Z, as discussed in Chapter 3. However, the B-, or beta-, axis movement (provided through the rotary index table, as discussed later in this chapter) provides the primary advantage of a horizontal machining center. The ability to perform a number of different operations such as milling, drilling, boring, counterboring, and tapping in a single workpiece setup while changing its own cutting tools is what has placed the horizontal machining center in the limelight.

Selection of either a vertical or a horizontal machining center depends mainly on the part type, size, weight, application, and, in many cases, personal preference. Each has its own specific advantages and disadvantages, as shown in table 8-1.

Some machine tool vendors now offer vertical/horizontal spindles (figure 8-2). These are similar in appearance to right-angle spindle attachments that have long been available to change the spindle orientation by 90°. Machines that can quickly convert themselves from vertical to horizontal or horizontal to vertical orientation continue to gain acceptance and prominence because they decrease the non–value-added setup and piece-handling time and increase the value-added chip-making time.

Five-axis CNC laser machining centers (see figure 8-3) are used for precision cutting, trimming, drilling, welding, and heat treating. Laser machines can handle most metals and alloys—including tool steel, tungsten, and titanium—as well as plastics, ceramics, and composite materials. Laser machines generate less heat than conventional cutters or saws, thereby minimizing part distortion, and exert no mechanical force that might stress the workpiece. Laser machining

TABLE 8-1 **General advantages and disadvantages of vertical and horizontal machining centers**

Vertical machining centers		Horizontal machining centers	
Advantages	**Disadvantages**	**Advantages**	**Disadvantages**
• Generally less costly.	• Not suitable for large, boxy, heavy parts.	• More flexible overall.	• Generally more costly.
• Thrust is absorbed directly into machine table during deep tool thrust operations such as drilling.	• As workpiece size increases, it becomes more difficult to conveniently look down into the cut.	• Table indexing capability enables multiple sides of a workpiece to be machined in one setting and clamping.	• Difficult to load and unload large, flat, plate-type parts.
• Ideal for large flat plate work, moldwork, and other single-surface three-axis contouring.	• Extensive chip buildup obstructs view of the cut and recuts chips.	• Chips drop out of the way during machining, providing an uncluttered view of the cut and preventing recutting of chips.	• High thrust must be absorbed by tombstones, fixtures, or right-angle braces.
• Heavy tools can be used without much concern about deflection.	• On large verticals, head weights and distance from the column can cause head drop, loss of accuracy, and chatter.	• Operator's station is to one side of the column, providing good line-of-sight control.	• Heavy tools can deflect.
		• Pallet shuttle exchange mechanisms are open, accessible, and easy to service.	
		• Ideally suited for large, boxy, heavy parts.	

FIGURE 8-2
Machining center that has both vertical and horizontal spindles and can programatically convert itself. (Courtesy of Maho Machine Tool Corp.)

centers can perform more efficiently and precisely than conventional machine tools. Consistent quality is easy to maintain because lasers suffer no tool wear, eliminating the need for tool-wear compensation.

Overall machining center construction has improved to accommodate higher spindle speeds, feeds, and horsepower capabilities, along with overall higher utilization rates and increased performance requirements. Machine beds, for the most part, are still made of the more traditional cast-iron or welded-steel plating. However, computer modeling of the final structures, using such techniques as finite-element analysis, has become more widespread. As a result, castings have been optimized to yield fewer distortions when the machine is in a load or cut condition.

Spindle head improvements have advanced to accommodate a fifth axis (and more) of movement, which greatly enhances a machining center's versatility. Key to the new designs are pitch and roll motions right in the spindle head. The additional axes of movement are a necessity, particularly on larger machining centers, where the workpiece cannot easily be moved and the tool must tilt and pivot to machine the stationary clamped part. But spindle heads that move in more than three primary axes are also becoming more popular for smaller workpieces. Parts previously machined in several settings on a ram-type universal milling machine, for instance, are moving over to small machining centers to be completed in one part setting.

Characteristics demanded of machining center spindles by modern high-performance cutting tool materials include stiffness, running accuracy (runout),

FIGURE 8-3
CNC laser machining center. (Courtesy of Cincinnati Milacron Inc.)

axial load-carrying capacity, thermal stability, and axial freedom for thermal expansion. Most importantly, though, the demand is for speed. In some cases, spindle speeds are in the 10,000- to 12,000-RPM range, depending on the manufacturer and the application required.

CNC machining centers represent a real frontier for future development. The unlimited potential of multiaxis capabilities will provide new and better ways to locate and machine various types of workpieces in the future.

TOOL STORAGE CAPACITIES

As already mentioned, one advantage of machining centers is that they automatically change their own cutting tools. By today's definition, a machining center must include an automatic tool changer. Tool storage and tool-changing mechanisms vary among the diversified machine tool suppliers; some are front-, side-, or top-mounted. This applies to both vertical and horizontal applications;

the tool magazine may have a vertical or horizontal axis. The tool magazine may then rotate so that the center of the tool is automatically aligned with the spindle. In most cases, the tool magazine is to one side or above the spindle. Figure 8-4 illustrates a vertical tool magazine and changing mechanism. Figure 8-5 shows a horizontal tool magazine and changing mechanism.

The advantages of having tools stored away from the working spindle include less contamination from flying chips and better protection for an operator changing tools during machining. The double-ended 180° indexing arm (figure 8-5) continues to be the most popular approach, although various designs of the tool gripping and clamping mechanisms will vary among builders.

Demand has increased for modern machining centers to have more cutting tools, which means a greater tool storage capacity is required. Machining requirements for cells, for example, demand that backup tools be available on-line to replace a broken or worn-out tool. Tools stored at machining centers fit into individual pockets of a machine tool's magazine, or tool matrix. Pocket designs vary, ranging from simple holes cut into a disk-shaped carousel to individually machined pockets assembled into a chain to interconnected plastic pockets.

The tools may be selected sequentially (in order of use or succession) or at random. Numbers may then be assigned to the position or pocket of the tool. This means that the programmer must, in some cases, decide which pockets or tool positions to use. The programmer also must supply the setup person with

FIGURE 8-4
Vertical tool magazine and changing mechanism. (Courtesy of Cincinnati Milacron Inc.)

FIGURE 8-5
Horizonatal tool magazine and changing mechanism. (Courtesy of Cincinnati Milacron Inc.)

a tool list identifying which cutting tool is to be used in each position. Certain tools may be assigned permanent positions in the tool magazine, depending on their frequency of use. This leaves the remaining pockets available for special tools.

The actual tool magazine for most machining centers and tool changers has holes around the perimeter of a circle, or on a chain, spaced at a specific distance. The maximum diameter of any cutter held in magazine storage cannot be greater than the hole spacing if all pockets are to be filled.

Tool-changing mechanisms vary greatly. Some have a combined operation of four machine elements to change tools:

- The *tool drum* contains 30 or more coded drum stations. It rotates in the direction commanded by the control.

- The *intermediate transfer arm* removes tools from the tool drum and places them in the interchange arm and later returns them to the tool drum.

- The *interchange station arm* receives tools from the intermediate transfer arm and swings them forward into the proper position for a tool interchange with the tool changer arm. It then swings them back into position for removal by the intermediate transfer arm.

- The *tool changer arm* simultaneously removes a tool from the interchange station arm and a tool from the spindle. It then interchanges these tools, inserting one into the spindle nose and the other into the interchange station arm.

One of the most important considerations for tool changers, regardless of the type, is whether the tool being removed from the spindle or the new tool going into the spindle will clear the workpiece and any other obstructions such as clamps and pushers. When programming with longer tools, the programmer must be extremely careful that the workpiece is moved far enough away from the spindle so that no collision will occur when the longer tool is inserted into the spindle. Often a retract to the extreme rear position in Z or an offsetting move in X or Y is required to avoid collisions.

TOOL LENGTH STORAGE/COMPENSATION

Tool length storage and *tool length compensation* allow the control to store information about a given tool length. This stored length is then applied to the Z-axis position when the tool is loaded into the spindle. This helps the programmer to program without knowing exact tool lengths and helps eliminate errors in calculating both the Z slide and tool length.

Practice has shown that the programmer must supply the operator with certain agreed-upon information to set up a job. With tool length storage/compensation, the following methods have worked very well.

Method 1

a) The programmer makes a tool assembly drawing for each tool used, showing all the components of a tool assembly (see figure 8-6). The tool set length is calculated and rounded off to the nearest one-eighth of an inch. When the tool is set up, the actual set length must be within a tolerance of the length dictated by the programmer.

b) The position where tool length values will be established by the programmer and the distance from this point to the centerline of index is determined. This distance and the feeler gage thickness are added to define a tool tram surface value. The operator enters this value, assigned by the programmer, into the control before setting tool length values.

c) After the operator loads the tools into the tool drum, each tool is located in the spindle and touched up to the tool tram surface. On some controls, pressing the TOOL LENGTH COMP. SET push button will initiate the calculation of a tool length and will store that information under the data for the tool.

Method 2

a) The programmer makes a tool assembly drawing for each tool used (figure 8-6), showing all the components of the tool assembly. The tool set length is calculated and rounded off to the nearest one-eighth of an inch. When the tool is set up, the actual set length must be within a tolerance of the length dictated by the programmer.

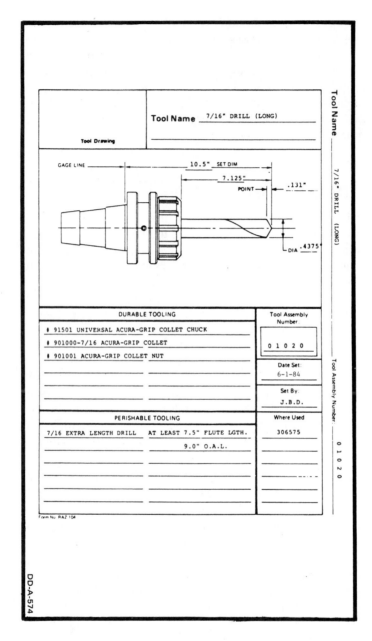

FIGURE 8-6
A typical tool assembly drawing. (Courtesy of Cincinnati Milacron Inc.)

b) The operator or tool specialist gages the exact length of each tool by means of a tool preset gage and makes note of it.

c) The operator loads tools into the tool drum and enters the length values for the tools into the control under their respective tool data locations.

Tool setting is not difficult because there are several simple mechanical and optical tool-setting devices. Sometimes the tools and toolholders are set by the toolroom, and the machine operator, having the time and ability, makes up the preset tools.

Many shops maintain tool assembly drawings. These drawings show the cutter, tool number, and setting distance. They simplify tool setting and provide consistent accuracy for all tools assembled. For example, every time a 10.375-in. set length, 0.500-in. diameter drill is used, the part programmer can check the drawing and find that a drill with those dimensions has a particular tool assembly number associated with it. The part programmer then calls for this tool assembly number in the part program. With this information, all part programmers can write programs calling for various tool assembly numbers and feel confident that everyone has the same dimensions.

WORK TABLES

Rotary index tables represent another feature that provides versatility for horizontal N/C machining centers. With proper fixturing and one clamping of the workpiece, the entire part can be machined in one setting. Rotary table motion is usually designated *B-axis*, or *β-axis*, and must be aligned the same as the X-, Y-, and Z-axes. An example of a rotary index table, showing the orientation of all axes, is given in figure 8-7.

Some rotary index tables are equipped with a universal fixture base and right-angle plate. These options maximize the productivity of a horizontal machining center. A conventional index table is shown in figure 8-8.

There are two basic types of rotary tables. However, the possibility of different methods of operation and control provide many possible combinations. The first type of rotary table uses a serrated plate to position the table mechanically. This rotary table will lift before indexing and lower into its final position after indexing. The second type of rotary table uses a rotary inductosyn seal to position and provide some means of feed rate control. This type of table may even be interpolated with the other slides to provide four-axis contouring.

Most rotary tables are bidirectional and will index using the shortest path to any of 72, 360, or 720 positions. A few of the different degrees of rotation in a rotary index table are illustrated in figure 8-9. The input is in degrees and, in most cases, all positions are in absolute positioning. Some rotary tables can index up to 360,000 positions. These tables are programmable in either absolute or incremental modes. Methods of programming sometimes vary with each manufacturer.

As much as rotary index tables can increase a machine tool's versatility and productivity, they can also be a real danger if the following considerations are not met:

1. The programmer should always make sure the tool is located in a safe position so that the table will index without hitting the cutting tool.

2. When a 180° rotation is required, two blocks of information should be programmed to ensure the desired direction of table rotation if the direction of rotation has a possible interference.

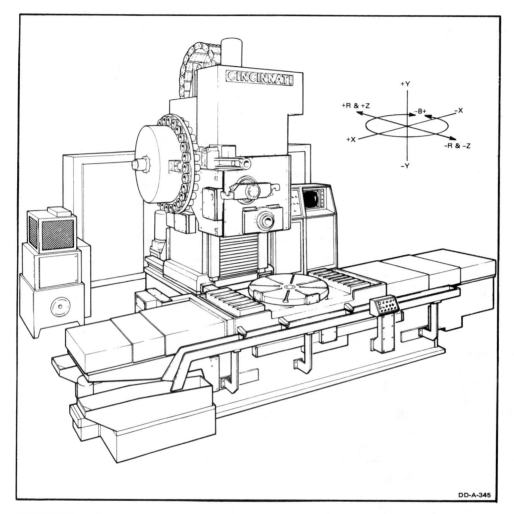

FIGURE 8-7
Rotary index table and machine, with axis orientation shown. (Courtesy of Cincinnati Milacron Inc.)

A few facts, and some tooling tips, are worth remembering and repeating. A machining center can put more different tools into the workpiece in a specified time period than any conventional machine tool. Thus, total tool use is generally greater. With so many cutting tools being applied without guidance of bushings, the cutting tools must have symmetry. Keep in mind that with all the precision built into the machine, it is no more accurate than the cutting tools it uses to machine the workpiece.

FORMAT INFORMATION

CNCs for machining centers accept decimal point programming or the interchangeable/word address format with either EIA or ASCII coding. Most con-

FIGURE 8-8
Conventional index table with part mounted on machine table. (Courtesy of Monarch
Machine Tool Company)

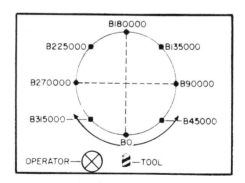

FIGURE 8-9
Rotary index table, illustrating
degrees of rotation.

trols will automatically sense which coding method is used (EIA or ASCII) and
will decode the N/C input accordingly.

The following list explains words used for a typical CNC machining center.
Some words are standard for any CNC machining center, whereas others may
be EIA- or manufacturer-specific. However, not all of these words are used for
every machining center and its programs. Nevertheless, the following is a brief
explanation of each word, the meaning of each character in the word, and their
use as they appear in a typical machining center tape format.

N Sequence number coding. It is a five-character code: one letter and four numerals (Oxxxx or Nxxxx). It is used to identify a block of information. It is informational rather than functional.

G Preparatory function coding. It is a three-character code: one letter and two numerals (Gxx). It is used for control of the machine. It is a command that determines the mode of operation of the system. This word is informational and functional. Therefore, all characters in this particular format, except leading zeros, must be included in the code.

X X-axis coordinate information code. It may contain up to nine characters: one letter, one sign, and up to seven numerals (X±xxx.xxxx inches or X±xxxx.xxx millimeters). Six numerals are normally used, though. This word is used to control the direction of table travel and position.

Y Y-axis coordinate information coding. This is identical to the X word, except that the Y address is used.

Z Z-axis coordinate information coding. This is identical to the X word, except that the Z address is used.

R Z-axis coordinate information coding. It may contain up to nine characters: one letter, one sign, and up to seven numerals (R±xxx.xxxx inches, R ±xxxx.xxx millimeters). It is used to control the positions of the Z slide at rapid traverse during positioning.

I The center point coordinate along the X-axis for circular interpolation. It may contain up to nine characters in the coding: one letter, one sign, and up to seven numerals (I±xxx.xxxx inches, I±xxxx.xxx millimeters).

J The center point coordinate along the Y-axis for circular interpolation. This is identical to the I word, except that the J address is used.

K The center point coordinate along the Z-axis for circular interpolation. This is also identical to the I word, except that the K address is used.

B Beta-axis coding of the rotary index table. This is a seven-character code: one letter and six numerals (Bxxx.xxx). It determines the angular position of the index table.

F Feed rate coding for X-, Y-, and/or Z-axes. It may contain up to five characters in the code: one letter and up to four numerals (Fxxx.x inches/minute, Fxxxx millimeters/minute). It is used for controlling the rate of longitudinal, vertical, and cross travel. The F word is also used in conjunction with the G04 code to dwell the slides. In this mode, the format can vary from 0.01 to 99.99 sec. of dwell.

S Spindle-speed coding for rate of rotation of the cutting tool. It may contain up to five characters in the code: one letter and up to four numerals (Sxxxx). This is the actual RPM desired for cutting.

T Tool number coding. It is a nine-character code: one letter and eight numerals (Txxxxxxxx). It determines the next tool to be used.

M Miscellaneous function coding. It is a three-character code: one letter and two numerals (Mxx). It is used for various discrete machine functions.

OPERATIONS PERFORMED BY MACHINING CENTERS

Many different types of operations are performed repeatedly by machining centers. Except for milling (which utilizes G01, discussed earlier in the text), many machining center operations use canned cycles in either the decimal point or interchangeable/word address programming format. Therefore, it is important to examine and thoroughly understand what these basic operations performed on a machining center are and how a typical machining center format appears.

In Chapter 5, the discussion of canned cycles involved the use of the decimal point programming format. In order to give students exposure to both primary formats in use today (decimal point programming and the interchangeable/word address format), the interchangeable/word address format will be used for the review of some machining center canned cycles.

Drilling

Drilling is an old, reliable method of metal removal, regardless of what machine is used to perform the operation. However, some helpful hints are worth mentioning.

Use the shortest drill possible to accomplish the job. Shorter drills are more rigid and are capable of greater accuracy. Lip height, clearance, and angle of point must be accurately ground for best results.

The following axis movements will occur when a G81 drill cycle is programmed:

1. Rapid in X and Y
2. Rapid in Z to gage height
3. Feed in Z to depth
4. Rapid retract to gage height

The use of the G81 drill cycle is illustrated in figures 8-10a, 8-10b, and 8-10c. Figure 8-10a shows the actions included in the G81 cycle. Figure 8-10b shows a sample program that uses the G81 code, and figure 8-10c shows the actions that this program accomplishes.

In sequence number O15 of the program, the G81 code rapid advances the tool along the X- and/or Y-axes simultaneously to position 1 from the previous position. When position 1 is reached, the tool rapids along the Z-axis to the R10.0000 plane (gage height). At this point, the tool feeds to the programmed Z depth at the programmed rate.

After reaching depth, the tool rapid retracts to the R10.0000 plane (gage height), and the next block of information (N16) is read and acted upon.

In sequence number N16, the G81 code is used again to rapid along the Y-axis, with the tool at gage height, to position 2. Then the tool feeds to the programmed Z depth at the programmed rate. After reaching depth* the tool rapid retracts to the R10.0000 plane (gage height).

---

*Programmed depth for the Z-axis is calculated as follows:

Z position = Depth of cut + Drill point = $1.0000 + (0.3 \times$ Diameter of drill)

= $1.0000 + (0.3 \times 0.5000) = 1.1500$

(The value of 0.3 is used for a standard 118° drill point.)

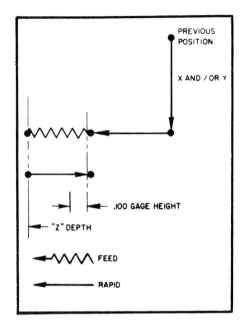

FIGURE 8-10a
G81: canned cycle drill schematic.

Boring

Boring is one of the most accurate ways to finish a hole. When starting with a drilled hole, the sequence of operations is usually to semifinish and finish bore. Starting with a cored hole, the operations generally required are rough, semifinish, and finish bore. Better boring will be achieved if the programmer makes sure that the following are true:

- The largest boring bar that will fit the hole to be machined is used.
- A chamfer tool is used, instead of a tool with a square shoulder, whenever possible.
- Multiple tool bars are used. The cutting operation should be planned so that the front cutter is through the work before the succeeding cutters start. This is because chatter from one cutter can be transmitted through the bar to the remaining cutters.
- Contour milling is employed, whenever practical, because it is often possible to eliminate the rough and semifinish operation by contour milling.

O/N SEQ.	G PREP. FUNCT.	X± POSITION	Y± POSITION	Z± POSITION	R± POSITION	I/J/K POSITION	A/B/C POSITION P/Q ± WORD		F/E FEED RATE	S SPINDLE SPEED	D WORD	T TOOL WORD	M MISC. FUNCT.
Ø 15	G81	X+ 40000	Y+100000	Z-11500	R+100000		B 0		F 100	S550		T 3	M03
N 16			Y 80000										

FIGURE 8-10b
G81: sample program.

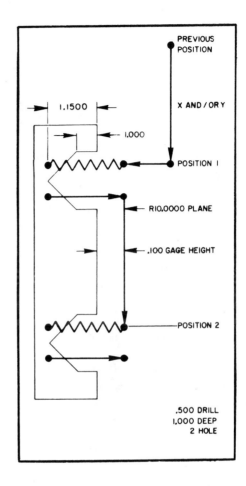

FIGURE 8-10c
G81: canned cycle drill—cycle representation.

The following axis movements will occur when a G85 bore cycle is programmed:

1. Rapid in X and/or Y
2. Rapid in Z to gage height
3. Feed in Z to depth
4. Feed retract to gage height

These four steps will occur in the same order every time a G85 cycle is programmed. Figures 8-11a, 8-11b, and 8-11c illustrate the use of a G85 bore cycle. Figure 8-11a shows the basic cycle, figure 8-11b shows a sample program that uses this cycle, and figure 8-11c shows the actions that this program accomplishes.

In sequence O15 of the program, the G85 code rapid advances the tool along the X- and/or Y-axes simultaneously to position 1 from the previous position. When position 1 is reached, the tool rapids along the Z-axis to the R10.0000 plane (gage height). At this point the tool feeds to the programmed Z

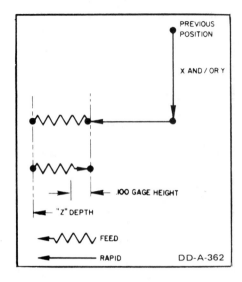

FIGURE 8-11a
G85: canned bore cycle schematic.

depth at the programmed feed rate. After reaching depth, the tool feed retracts to the R plane (gage height), and the next block of information (N16) is read and acted upon.

In sequence N16, the G85 code is used again to rapid along the Y- axis, with the tool at gage height, to position 2. The tool then feeds to the programmed depth at the programmed rate. After reaching depth, the tool feed retracts to the R plane (gage height).

Tapping

When programming tapping operations, be sure that the proper drill has been specified or that the bored hole size is correct. All taps should have adequate clearance to provide chip disposal. Avoid using straight-fluted hand or machine taps except when tapping a material such as cast iron. Chips can "ball up" during entry. This may break the tap during "backout." Spiral-fluted taps are better, especially for blind or deep holes, because the spiral causes the chips to feed up the length of the tap and out the hole. For through holes no longer than twice the tap diameter, a straight flute, spiral point, or "gun" tap can be used. This tap has a negative lead ground on the start, or chamfer end,

O/N SEQ	G PREP. FUNCT.	X± POSITION	Y± POSITION	Z± POSITION	R± POSITION	I/J/K POSITION	A/B/C POSITION P/Q ± WORD		F/E FEED RATE	S SPINDLE SPEED	D WORD	T TOOL WORD	M MISC. FUNCT.
Ø 15	G85	X+ 40000	Y+ 100000	Z− 10500	R+ 100000		B 0		F 40	S 623		T 9	M02
N 16			Y 80000										

FIGURE 8-11b
G85: sample program.

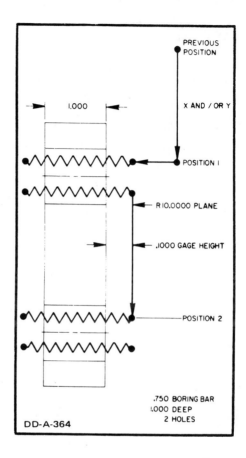

PREVIOUS
POSITION

X AND / OR Y

1.000

POSITION I

R10.0000 PLANE

.1000 GAGE HEIGHT

POSITION 2

.750 BORING BAR
1.000 DEEP
2 HOLES

DD-A-364

FIGURE 8-11c
G85: canned cycle bore—cycle representation.

that causes chips to be thrown ahead of the tap. At reversal, it leaves the chips there. Chip difficulties can be eliminated in some cases by the use of fluteless taps that roll or form the thread into the walls of the hole. In summary, use gun taps for short through holes, spiral-fluted taps for deep or blind holes, and fluteless taps whenever possible.

The following axis movements will occur when a G84 tap cycle is programmed:

1. Rapid in X and/or Y
2. Rapid in Z to gage height
3. Feed in Z to depth
4. Reverse spindle direction of rotation, and feed retract along the Z-axis to gage height
5. Reverse the spindle again at gage height

These five steps will occur in the same order every time a G84 cycle is programmed.

NOTE: If a right-hand thread is to be tapped, the M function should be for a CW (clockwise) spindle rotation. If a left-hand thread is to be tapped, the M function should be for a CCW (counterclockwise) spindle rotation.

Figures 8-12a, 8-12b, and 8-12c illustrate the use of a G84 tap cycle. Figure 8-12a shows the basic cycle, figure 8-12b shows a sample program that uses this cycle, and figure 8-12c shows the actions that this program accomplishes. In sequence O15 of the program, the G84 code rapid advances the tool along the X- and/or Y-axes simultaneously to position 1 from the previous position. When position 1 is reached, the tool rapids to the R10.0000 plane (gage height). At this point, the tool feeds to the programmed Z depth* at the programmed rate, with the spindle rotating in its primary direction as specified by the M function. At depth, the spindle reverses direction of rotation and feed retracts to the R10.0000 plane (gage height). Spindle rotation reverses again to the primary direction. The next block of information (N16) is then read and acted upon.

In sequence N16, the G84 code is used again to rapid along the Y-axis, with the tool at gage height, to position 2. Then the tool feeds to the programmed depth as before. At depth, the spindle reverses direction of rotation and feed

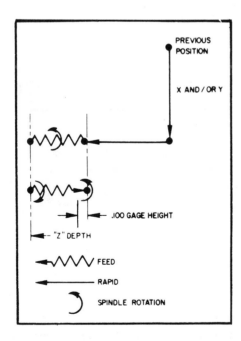

FIGURE 8-12a
G84: canned cycle tap schematic.

---

*The programmed depth for the Z-axis in this example is calculated as follows:

$$Z \text{ tapping depth} + \left[\text{Tap chamfer} \times \frac{1}{\text{Pitch}} \right] - \left[\text{Revolutions for reversal} \times \frac{1}{\text{Pitch}} \right]$$

This example is based on a $\frac{1}{4}$-20 tpi (threads per inch) tap with three-thread chamfer. Pitch is 20. Therefore, (1/Pitch) = (1/20) = 0.050 in.

O/N SEQ	G PREP FUNCT	X± POSITION	Y± POSITION	Z± POSITION	R± POSITION	I/J/K POSITION	A/B/C POSITION P/Q ± WORD		F/E FEED RATE	S SPINDLE SPEED	D WORD	T TOOL WORD	M MISC. FUNCT.
Ø 15	G84	X+ 40000	Y+ 100000	Z− 8470	R+ 100000		B 0		F 300	S 700		T 15	M03
N 16			Y 80000										

FIGURE 8-12b
G84: sample program.

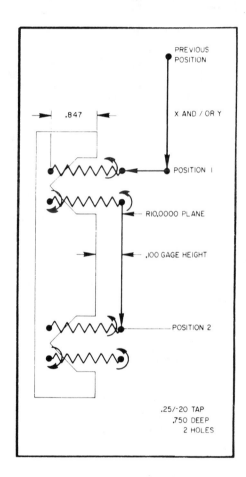

FIGURE 8-12c
G84: fixed cycle tap—cycle representation.

retracts to the R10.0000 plane (gage height). Spindle rotation then reverses again back to the primary direction.

PALLETS, PART LOADING, AND PROGRAMMING OPTIONS

Pallets are basically movable machine tool tables, as shown in figure 8-13. They enable the operator to set up, load, and unload parts and fixtures off-line while

FIGURE 8-13
Pallets (movable machine tables) help maximize spindle utilization time by enabling off-line part setup, load, and unload. (Courtesy of Cincinnati Milacron Inc.)

another part-loaded pallet is being machined. In a palletized system, the spindle is dedicated to machining because the operator sets up workpieces away from the machine. An automated pallet transfer device may be used to transfer material to and from a load/unload queue or among other machine tools in a cell or system application.

Pallet use has increased through the development of flexible manufacturing cells and systems, although pallets are increasingly being used in stand-alone machine tool applications with automatic work changers attached to the front of the machine (see figure 8-14). Pallets have been applied to vertical and horizontal machining centers, vertical turret lathes, grinders, wash stations, inspection machines, coordinate measuring machines, and other specialized equipment.

Elements of a pallet system are as follows:

1. The pallet, upon which is mounted either the workpiece or the fixture holding the workpiece.
2. The receiver, which is mounted on the machine tool table; it grips and holds the pallet.
3. The loader, which passes the pallet from the load station or queuing carousel to the receiver (stand-alone machine tool) or passes the transport carrier to the load station (cell or system). Loading and clamping are usually hydraulically actuated, and the pallet is located on the receiver by means of shot pins forced from the receiver into corresponding bushings on the pallet to ensure precision positioning.

The use of pallets in a cell, system, or stand-alone machine tool application offers several advantages:

- *Increased uptime at the spindle.* Machine tool utilization rates have increased 50 to 100% in many cases because the spindle does not have to stop while the next job is being set up.
- *Reduced total part setup time.* Some workpieces are best done with some of the operations completed on one machine and the remainder on an-

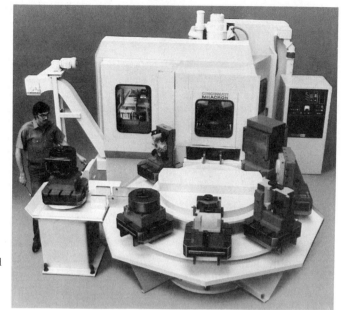

FIGURE 8-14
An automatic work changer equipped with an off-line load/unload station and an eight-pallet queue for horizontal machining center part processing. (Courtesy of Cincinnati Milacron Inc.)

other. By putting one setup on a pallet and then moving the palletized workpiece from one machine to the next, the total setup time is reduced and utilization of both machines is increased.

- *Improved quality.* Reducing multiple setups reduces the risk of introducing quality problems that can result from movement to a new location, reclamping, and distortion problems. In a palletized approach, the operator sets up a workpiece on a pallet once, before the pallet is engaged at the first machine tool. After that, if other machining operations are required and an automated transfer system is used, as shown in figure 8-15, the pallet is transferred to the second and third machines and then to an unload area or possibly to another cell.

Part- and pallet-loading schemes related to stand-alone, cell, and FMS applications vary considerably among users. Various types of fixturing and tombstoning are used to configure and locate prismatic parts for machining center part processing.

Use of standard tombstoning, which allows parts to be mounted on a tombstone (see figure 8-16) and machined in one setup, is increasing because of the surge in popularity of palletized machining. Parts may be mounted to a standard four-sided tombstone for horizontal machining center processing through one of three conventional schemes:

1. *Single-load mounting* (see figure 8-17a). One part is loaded on one side of a four-sided tombstone, providing access to only one surface at a time without indexing the rotary B-axis. This scheme is highly inefficient and does not take full advantage of the tombstone's remaining three sides.

FIGURE 8-15
Pallets can be loaded and unloaded off-line and quickly transferred from machine to machine in an automated machining cell or system. (Courtesy of Cincinnati Milacron Inc.)

FIGURE 8-16
Tombstone bases are used for clamping parts for horizontal machining centers. (Courtesy of Cincinnati Milacron Inc.)

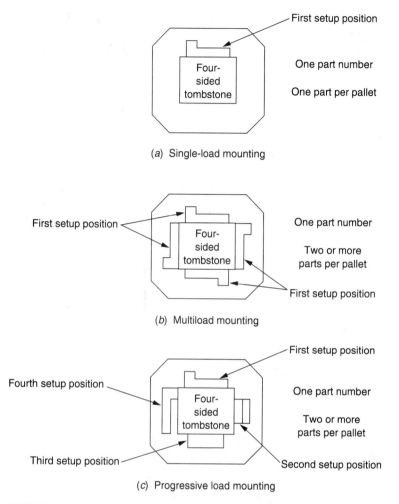

(a) Single-load mounting

(b) Multiload mounting

(c) Progressive load mounting

FIGURE 8-17
The three conventional pallet/part-loading schemes generally used with four-sided tombstones.

2. *Multiload mounting* (see figure 8-17b). Multiple parts are loaded on two, three, or four sides of a tombstone. This scheme is very efficient and takes full advantage of the tombstone's four sides. The same surface of multiple parts is presented for machining at each B-axis index because the tombstone is loaded with parts all mounted identically. Once the machining of similar surfaces for the lot of parts has been completed, subsequent part settings exposing different surfaces of the same part can be multiloaded on the same tombstone for machining. Or these subsequent settings can be mounted on a different tombstone, pallet, and machine (if available) while the original machining center machines another set of parts.

3. *Progressive load mounting* (see figure 8-17c). Progressive loading also takes full advantage of the tombstone's four sides. It involves progressively moving the part or parts through different settings, exposing all sides of the workpiece and removing a completed part at the end of each programmed machining cycle.

Another loading scheme used by some manufacturers involves manufacturing parts by ship sets. Ship set manufacturing consists of fixturing all applicable parts required to build one complete unit or subassembly on an entire automatic work changer (see figure 8-18). Thus, one complete cycle of an eight-to-ten-pallet automatic work changer (six to eight hours total machining time, for example) could yield all noncommercial parts required to build one particular subassembly. The entire ship set of parts could then be sent to assembly, as opposed to machining multiple individual component parts in batches.

Selection of a tombstone, pallet, or part-loading scheme will vary considerably among machine tool users, depending on the application of the machine tool (stand-alone, cell, or FMS) and other factors such as part type, size and weight, accuracies to be maintained, average lot size, and machine and tooling capability and capacity.

FIGURE 8-18
An example of ship set manufacturing fixtured on an eight-pallet queuing carousel for horizontal machining center part processing. (Courtesy of Cincinnati Milacron Inc.)

Understanding programming options is important to part-processing cycle time and overall machining efficiency, regardless of whether one of the processor languages (such as APT or an interactive graphics-based programming package) is used. Programming option selection refers to the particular method and approach of cutting tool use for the processing of multiple parts in one setup at each machine.

Programming options for processing multiple parts on a four-sided tombstone, for example, can be either tool dominant or part dominant. *Tool-dominant* programming implies using each cutting tool at every part location requiring that specific tool, on all four sides of the tombstone, before exchanging that cutting tool for another. Indexing the part or tombstone is generally less time consuming and more efficient than changing to the same tool several times during the program.

Part-dominant programming implies using and exchanging all the cutting tools to complete a particular part or tombstone side before indexing to another part or side of the tombstone. To some programmers, exchanging cutting tools, possibly several times on one tombstone side, may be less time consuming and more efficient than indexing the part or tombstone and using the same tool in all required locations before exchanging it for another.

Neither tool- nor part-dominant programming is better or more efficient in all cases. In some circumstances, because of the type, variety, and number of cutting tools required for specific parts, both are used within the same program. And, because of the random tooling feature available on modern machining centers, part programmers can call for a tool change whenever desired. However, machining efficiencies may increase or decrease, depending on whether tool- or part-dominant programming is used. Critical methods engineering analysis should be employed to study the particular part, setup, and cutting tool application and to determine the best techniques for optimal part-processing efficiency.

AUTOMATED FEATURES AND CAPABILITIES

Three principal developments in the 1960s led to increased acceptance and use of horizontal machining centers:

1. The capability of a machine tool to change its own cutting tools on command

2. An indexable work table permitting machining on multiple sides of a workpiece in one clamping (increased versatility of horizontal over vertical machining centers)

3. The ability to call up interchangeable pallets from an off-line bank for machining center part processing

Today's trend is to incorporate many diverse functions in a single machining center, such as tilt tables, swivel spindles, and touch-trigger probing. The result is the emergence of the flexible manufacturing cell resident in a single machine tool.

Machining center automated features and capabilities perform a multitude of functions that now automate what was previously performed manually in several separate operations and on a variety of different machine tools. The following subsections describe some of the principal automated machining center features and capabilities.

Probing

Probing, a creative and time-saving method of accomplishing work centering (see figure 8-19), is used to trigger electronically the programmed surface reference in X, Y, or Z and make automatic compensation for the measured axis values through direct feedback. This touch sensor tool can be used to automatically regrid the machine, locate the setup point, establish tool clearance planes, or determine a reference point for machining with respect to the center of a boss or cored hole.

The surface-sensing probe contains three parts: (1) the probe body with an interchangeable stylus (the stylus makes physical contact with the workpiece); (2) the noncontacting inductive modules (one module is mounted on the nose of the machine spindle, whereas another is mounted on the probe itself); and (3) the control and the interface printed circuit board.

The precision surface-sensing probe can be loaded and automatically selected from the tool storage matrix of machining centers, as shown in figure 8-20. Probing can improve the machining accuracy by feeding back offsets to fine-tune the program in the range of 0.0001 in. or finer. This technique bypasses the need for extremely fine (and costly) drives and position-measuring devices in the machine tool. Additionally, probing improves productivity because it

FIGURE 8-19
The probe: a precision surface-sensing tool. (Courtesy of Cincinnati Milacron Inc.)

FIGURE 8-20
The surface-sensing probe
loaded in the tool storage
matrix of a horizontal
machining center. (Courtesy
of Cincinnati Milacron Inc.)

centrally locates the finished part within the envelope of the rough part, ensuring part cleanup. The result is a significant, immediate reduction in time wasted by the machining of parts that do not have sufficient stock. There is also an immediate reduction in the time required to lay out parts manually by bluing and line scribing.

Probing is used in some cases for on-machine part inspection following completion of all machining operations. This can have both a positive and negative impact on operations. On-machine inspection will, in many cases, eliminate the high cost of a coordinate measuring machine (CMM). It can also eliminate the need to set up, load, and unload, and handle the part for inspection and the need to write and maintain a separate CMM inspection part program. Thus, the cost of having the part sit and wait for inspection (which adds another operation) can be eliminated, thereby decreasing work-in-process inventory and increasing part throughput. Additionally, inspection/machining conditional logic can be merged so that if more cuts are required during the inspection process, the program will automatically branch to the appropriate section in the part program to make another machining pass and reinspect. All this saves time and money for manufacturers.

On the other hand, on-machine inspection ties up a CNC machine tool for non–value-added inspection purposes; the tool could be adding value by performing another machining operation while a separate CMM performs the inspection. On-machine inspection also increases the size of the part program and the amount of time required to actually do the programming (both by as much as a third) by adding all the conditional branching and logic routines to the part program.

Broken-Tool Detection

Broken-tool detection, accomplished through the N/C part program, permits moving each tool to a fixed probe position to check for tool breakage before the cut sequence begins (see figure 8-21). If a tool is broken, the machine will automatically replace it with a duplicate stored in the tool matrix. If a duplicate does not exist, a machine-stop condition will occur and operator action is required. Broken-tool detection helps increase machine productivity and utilization and decrease operator involvement and attention.

Multiple-Spindle and Angled Spindle Heads

When cycle time can be improved by drilling or tapping several holes at once, multiple-spindle, or cluster heads may be used. Multiple-spindle heads (see figure 8-22), which can be loaded like an ordinary tool, drive a cluster of tools through their internal gearing mechanisms. The head design contains a fixed number of driven spindles, but the location of each spindle relative to the others is determined by a specific hole pattern.

Angle heads of 90° and 45° (see figure 8-23) are used on machining centers in highly specialized applications. Typically, they are used where the investment in an angled spindle head can be justified.

Automated Tool Delivery

Automated tool delivery to a machining center in a stand-alone mode or as part of an automated cell or system offers big gains in productivity and machine

FIGURE 8-21
Broken-tool detection checks for tool breakage before or after the cutting sequence. (Courtesy of Cincinnati Milacron Inc.)

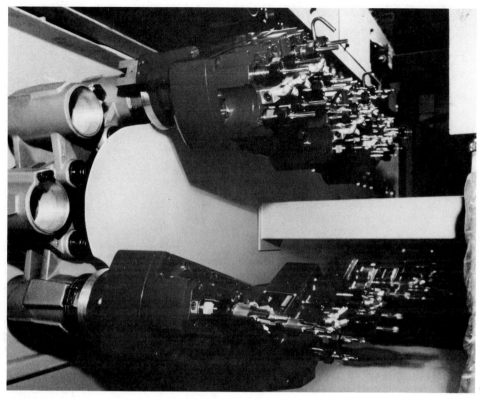

FIGURE 8-22
Multiple-spindle heads can be loaded like an ordinary tool and drive a cluster of tools to machine a repetitive hole pattern. (Courtesy of Cincinnati Milacron Inc.)

utilization rates because machines do not need to be stopped for tool replacement. Sometimes delivered on an automated guided vehicle (AGV) to the rear of the machine and tool matrix (see figure 8-24), new cutting tools can be exchanged with used tools without any interference with the ongoing machining process. Although sometimes controlled through the added help of a cell controller or minicomputer, the automated tool delivery and exchange capability add considerably to a machining center's overall uptime and performance.

Torque Control Machining

Sometimes referred to as adaptive control, torque control machining was developed to speed up or slow down a cutting tool while the tool is engaged in the actual cutting operation. The function of torque control machining is to sense machining conditions (see figure 8-25) and adjust the feeds and speeds to suit the real-time condition. Sensing devices are built into the machine spindle to sense torque, heat, and vibration. These sensing devices provide feedback

FIGURE 8-23
Angle heads of 90° and 45° are sometimes used in highly specialized machining applications.
(Courtesy of Cincinnati Milacron Inc.)

signals to the MCU, which contains the preprogrammed safe limits. If the pre-programmed safe limits are exceeded, the MCU alters the feeds and speeds, up or down, to suit the changing conditions.

Fixture Offsets

Fixture offsets are an important feature available on a variety of MCUs. Simply stated, they enable the operator to compensate for erroneous dimensional differences in the fixture, its location, or the part itself. Vitally important to horizontal machining centers where work is located on a rotary index table, the fixture offset feature permits the operator to enter the offset value for each axis involved; the MCU will automatically compensate for that difference through the entire part program. Fixture offsets can also be assigned exclusively to specific part features or specific operations within the part program.

Fixture offsets, (see figure 8-26), are most commonly used when an operator loads a fixture on the machine table and finds that the fixture is off-center. A one-time-only correction in the X-, Y-, or Z-axes through fixture offsets (the operator enters the correction difference into the MCU) enables machining on all sides of the workpiece as if the fixture was located on the centerline of index.

FIGURE 8-24
Automated tool delivery generally occurs at the rear of the machine and tool matrix and eliminates stopping the machine for tool replacement. (Courtesy of Kearny and Trecker Corp.)

FIGURE 8-25
Torque control machining senses machining conditions and adjusts the feeds and speeds, up or down, to suit the real-time cutting conditions of each tool. (Courtesy of Kennametal Inc.)

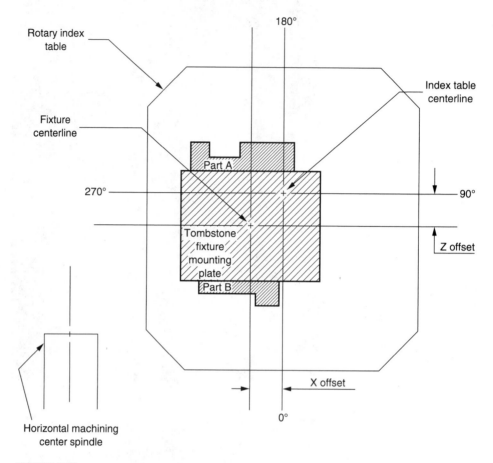

FIGURE 8-26
With fixture offsets, the table automatically "zero shifts" with every table index to compensate for X, Y, and Z fixture-off-center conditions.

Each time a rotary index is called for and completed, the table automatically zero shifts the required offset amount. Other examples where fixture offsets can be used include parts with features like a protruding boss to be circular milled or a cored hole to be drilled. In these examples, the relative position on the part may vary from lot to lot but not from piece to piece within the lot. The programmer can assign a fixture offset for the machining of that feature, and the operator can determine the positional-error difference and enter the offset value through the MCU console keyboard.

Fixture offset words begin with the word address H followed by two digits. The number of offsets can range up to 32 (H00–H32) or more. For each H-code word entered in the part program, separate offset values can be utilized for up to three axes. Fixture offset values are modal and remain in effect until a new H code is entered or the program is ended.

EXAMPLE PROGRAMS

The following example programs are typical of parts that require various tooling to perform some common machining center operations. In some cases, only certain excerpts have been included to avoid lengthy lists of repetitive operations.

Undoubtedly, there are approaches and proprietary techniques other than those detailed here. However, these examples should serve to better acquaint you with the basic functions and operations of modern machining centers.

The first program is for a vertical machining center. The block-by-block explanation helps explain the decimal point programming format used to complete the necessary machining. The final part for this example is shown in figure 8-27a.

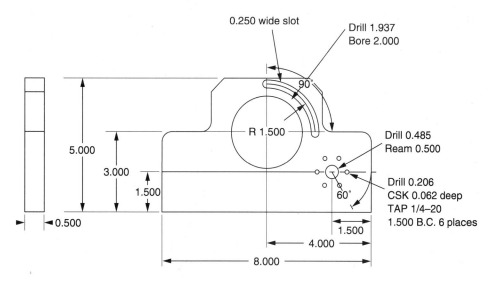

FIGURE 8-27a
Final part for vertical machining center program example (all units in inches).

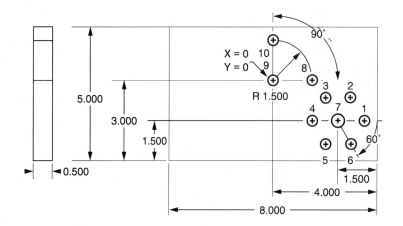

FIGURE 8-27b
Vertical machining
center example:
center drill holes.

Refer to figure 8-27b.

01000		Program number
N010	G54 G90 G80 G40 G00	Absolute input, cancel, rapid
N020	M6 T1	Tool change to tool 1 (center drill)
N030	M03 S1500	Turn on spindle, 1500 RPM
N040	G43 H1. Z.5	Tool length offset
N050	X3.25 Y-1.5	Rapid to position 1
N060	G81 G99 X3.25 Y-1.5 Z-.2R.05 F3.0	Canned cycle, Center drill hole 1
N070	X2.875 Y-0.85	Hole 2
N080	X2.125	Hole 3
N090	X1.75 Y-1.5	Hole 4
N100	X2.125 Y-2.15	Hole 5
N110	X2.875	Hole 6
N120	X2.5 Y-1.5	Hole 7
N130	X1.5 Y0.	Hole 8
N140	X0	Hole 9
N150	Y1.5	Hole 10
N160	G80	Cancel drill cycle G81
N170	G00 G28 G91 Z0	Rapid to tool change position
N180	M5	Turn off spindle

Refer to figure 8-27c.

N190	M6 T2	Tool change to tool 2 (0.206-in.dia. drill)
N200	M3 S850	Turn on spindle, 850 RPM
N210	G43 H2 Z.5	Tool length offset, tool 2
N220	X3.25 Y-1.5	Rapid to position 1
N230	G83 X3.25 Y-1.5 Z-.55 Q.1 R0.05 F3.0	Canned cycle, Peck drill hole 1
N240	X2.875 Y-0.85	Hole 2
N250	X2.125	Hole 3
N260	X1.75 Y-1.5	Hole 4
N270	X2.125 Y-2.15	Hole 5
N280	X2.875	Hole 6
N290	X1.5 Y0.	Hole 8
N300	X1.5	Hole 10
N310	G80	Cancel drill cycle G83

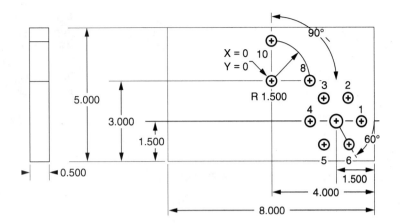

FIGURE 8-27c
Vertical machining center example: canned cycle peak drill tap holes.

N320	G28 G91 G0 Z0	Rapid to tool change position
N330	M5	Turn off spindle
N340	M6 T3	Tool change to tool 3 ($\frac{1}{4}$-20 tap)
N350	M3 S200	Turn on spindle, 200 RPM
N360	G43 H3 Z.5	Tool length offset
N370	G0 X3.25 Y-1.5	Rapid to position 1
N380	G84 R.1 Z-.55 F10.	Tap hole 1
N390	X2.875 Y-.85	Hole 2
N400	X2.125	Hole 3
N410	X1.75 Y-1.5	Hole 4
N420	X2.125 Y-2.15	Hole 5
N430	X2.875	Hole 6
N440	G91 G28 G0 Z0	Rapid to tool change position
N450	M5	Turn off spindle

Refer to figure 8-27*d*.

N460	M6 T4	Tool change to tool 4 (31/64 drill)
N470	M3 S450	Turn on spindle, 450 RPM

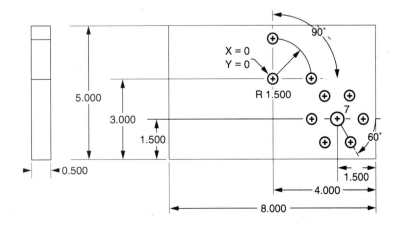

FIGURE 8-27d
Vertical machining center example: peck drill/bore hole.

N480	G43 H4 Z.5	Turn on Tool length offset
N490	G0 X2.5 Y-1.5	Rapid to position 7
N500	G83 X2.5 Y-1.5 Z-.550 R.05 Q.1 F3.0	Peck drill cycle
N510	G28 G91 G0 Z0	Rapid to tool change position
N520	M5	Turn off spindle
N530	M6 T5	Tool change to tool 5 (0.500in. ream)
N540	M3 S220	Turn on spindle, 220 RPM
N550	G43 H5 Z.5	Tool length offset
N560	G0 X2.5 Y-1.5	Rapid to position 7
N570	G85 X2.5 Y-1.5 Z-.55 R.1 F6.	Bore cycle, feed in, feed out
N580	G80	Cancel G85
N590	G0 G91 G28 Z2.0	Rapid to tool change position
N600	M5	Turn off spindle

Refer to figure 8-27e.

N610	M6 T6	Tool change to tool 6 (1.937-in. drill)
N620	M3 S120	Turn on spindle, 120 RPM
N630	G43 H6 Z.5	Tool length offset
N640	G0 X0. Y0.	Rapid to position 9
N650	G83 X0. Y0. Z-.55 R.1 Q.1 F3.	Peck drill cycle
N660	G80	Cancel G83
N670	G28 G91 G0 Z0	Rapid to tool change position
N680	M5	Turn off spindle
N690	M6 T7	Tool change to tool 7 (boring head)
N700	M3 S200	Turn on spindle, 200 RPM
N710	G0 X0 Y0	Rapid to position 9
N720	G43 H7 Z.5	Tool length offset
N730	G85 X0. Y0. Z-.55 R.1 F3.	Boring cycle, feed in, feed out
N740	G80	Cancel G85
N750	G0 G28 G91 Z0	Rapid to tool change position
N760	M5	Turn off spindle

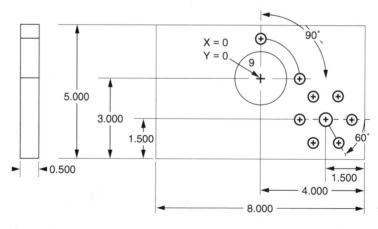

FIGURE 8-27e
Vertical machining center example: additional peck drilling; bore hole.

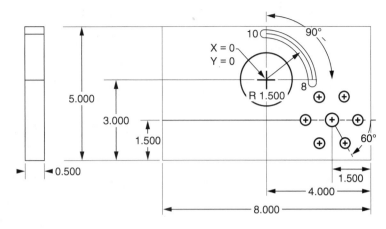

FIGURE 8-27f
Vertical machining center example: mill arc (using subprogram).

Refer to figure 8-27f.

N770	M6 T8	Tool change to tool 8 (0.187 in. end mill)
N780	M3 S850	Turn on spindle, 850 RPM
N790	G0 X0 Y1.5	Rapid to position 10
N800	G43 H7 Z.1	Tool length offset
N810	G01 Z0	Feed down to top of part
N820	M98 P1001 L5	Perform subprogram 1001 five times
N830	G90 G0 Z2.	Rapid 2.0 in. above part
N840	G28 G91 G0 Z0	Rapid to tool change position
N850	M5	Turn off spindle
N860	M6 T8	Tool change to tool 8 (0.250-in. end mill)
N870	M3 S800	Turn on spindle, 800 RPM
N880	G0 X0 Y1.5	Rapid to position 10
N890	G43 H8 Z.1	Tool length offset
N900	G1 Z-.55	Feed to depth of 0.550 in.
N910	G02 X1.5 Y0 J-1.5	Swing arc to end point 8
N920	G1 Z.1 F25.	Feed out 0.100 in. above part
N930	G28 G91 G0 Z0	Rapid to tool change position
N940	M5	Turn off spindle

Subprogram 1001

N821	G91 G1 Z-.055 F15.	Incrementally feed down 0.055 in.
N822	G90 G02 X1.5 Y0. J-1.5	Swing arc to point 8
N823	G91 Z-.055	Incrementally feed down 0.055 in.
N824	G90 G03 X0 Y1.5 I-1.5	Swing arc to point 10
N825	M99	Calls up main program
N950	M6 T1	Tool change to tool 1 (center drill)
N960	M30	Reset program

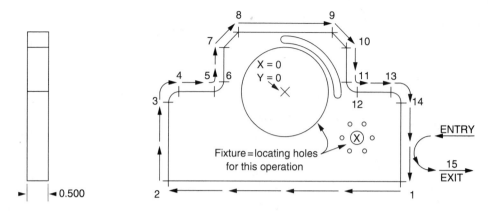

FIGURE 8-27g
Vertical machining center example: mill periphery of part (using subprogram).

For the remaining milling operation, the part will be held in a fixture locating off holes 7 and 9 (see figure 8-27b). Refer to figure 8-27g.

N961	G54 G90 G80 G40	
N962	M6 T9	Tool change to tool 9 (0.625-in. ROUGH)
N963	M3 S450	Turn on spindle, 450 RPM
N970	G0 X6.3 Y-1.375	Rapid to entry start point
N980	G43 H9 Z.5	Tool length offset
N990	G1 Z-.55 F15.	Feed down to depth of 0.550 in.
N1000	G1 G41 D59 X4.5 F10.	Cutter comp left; D59 dia. offset Add 0.030 in. to dia. offset (0.342 in.)
N1010	G03 X4. Y-1.875 R.5	Entry arc
N1020	M98 P1002 L1	Call up subprogram 1002 one time
N1030	G90 G1 Z.1	Feed above part 0.100in.
N1040	G28 G91 Z0	Rapid to tool change position
N1050	M5	Turn off spindle
N1060	M6 T10	Tool change to tool 10 (0.500-in. finish)
N1070	M3 S450	Turn on spindle, 450 RPM
N1080	G0 X6.3 Y-1.375	Rapid to entry start point
N1090	G43 H10 Z.5	Tool length offset
N1100	G1 Z-.55 F15.	Feed down to depth of 0.550 in.
N1110	G1 G41 D60 X4.5 F10.	Cutter comp left; D60 dia. offset Add 0 to dia. offset (0.250 in.)
N1120	G03 X4. Y-1.875 R.5	Entry arc
N1130	M98 P1002 L1	Call up subprogram 1002 one time
N1140	G1 G90 Z.1	Feed up 0.100 in. above part
N1150	G28 G91 Z0	Rapid to tool change position
N1160	M5	Turn off spindle
N1170	M6 T1	Tool change to tool 1 (center drill)
N1180	M30	Reset program back to beginning

Subprogram 1002

N1190	G1 X4. Y-3.	Feed to point 1
N1200	G01 X-4.	Point 2
N1210	Y-0.5	Point 3
N1220	G02 X-3.5 Y0. I0.5	Point 4
N1230	G01 X-2.5	Point 5
N1240	G03 X-2. Y0.5 J0.5	Point 6
N1250	G01 Y1.5	Point 7
N1260	X-1.5 Y2.	Point 8
N1270	X1.5	Point 9
N1280	X2. Y1.5	Point 10
N1290	Y0.5	Point 11
N1300	G03 X2.5 Y0. I0.5	Point 12
N1310	G01 X3.5	Point 13
N1320	G02 X4. Y-0.5 J-0.5	Point 14
N1330	G01 Y-2.375	Beginning of exit line arc
N1340	G03 X4.5 Y-2.875 R.5	Exit arc
N1350	G1 X6.	Point 15
N1360	M99	Call up main program

The second example involves a horizontal machining center using the interchangeable variable block format. The following information is provided:

1. Figure 8-28*a* shows an engineering drawing of a hydraulic pump housing.

2. Figure 8-28*b* shows a plan view and figure 8-28*c* an elevation view of the one-station indexing fixture. This fixture would permit machining on four sides of the part.

3. The beginning of the N/C program is shown in figure 8-28*d* along with a detailed explanation for this part of the program.

4. Figure 8-28*e* illustrates numbered hole positions on the 0-side of the part. Specific blocks of information for drilling the numbered hole positions have been omitted because of their repetitive nature.

5. Figure 8-28*f* includes blocks of information and a related explanation for a rotary table index from the *O'* to the 180-side of the part for the remaining machining operations.

6. Figure 8-28*g* includes blocks of information and a related description for drilling and tapping the hole positions illustrated in figure 8-28*h*.

7. The blocks of information in figure 8-28*i* provide the programming information required to profile mill a cored hole to a rough bore, as shown in figure 8-28*j*.

8. A tool instruction sheet, with complete tool identification and cutting statistics, is shown in figure 8-28*k*.

9. Figures 8-28*l* and 8-28*m* illustrate examples of tool drawings that document the various parts of the entire tool assembly and complete the programming support paperwork.

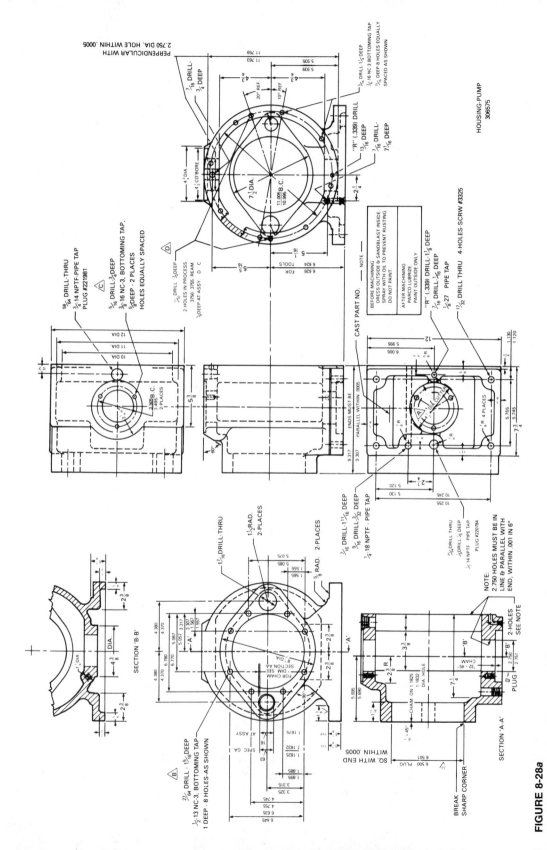

FIGURE 8-28a

Sample part drawing. (Courtesy of Cincinnati Milacron Inc.)

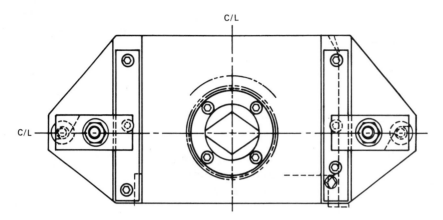

FIGURE 8-28*b*
Fixture plan view. (Courtesy of Cincinnati Milacron Inc.)

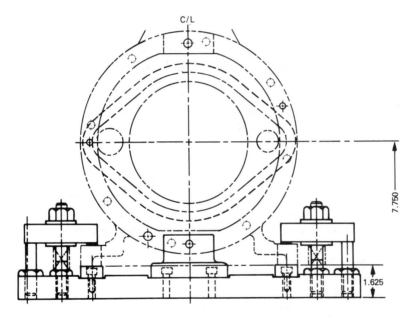

FIGURE 8-28*c*
Fixture elevation view. (Courtesy of Cincinnati Milacron Inc.)

O/N SEQ.	G PREP. FUNCT.	X ± POSITION	Y ± POSITION	Z ± POSITION	R ± POSITION	I/J/K POSITION	A/B/C POSITION P/Q ± WORD	F/E FEED RATE	S SPINDLE SPEED	D/H WORD	T TOOL WORD	M MISC. FUNCT.
0 1	G00										T 1	M 6
0 2	G00	X 75000	Y 77500	Z 56925			B 0		S 350		T 2	M 3

Tape Sequence:
(E.O.B.)

O1   Tape alignment blocks are identified by the alphabetic character "O" as per E.I.A. standards. With tool changing machining centers, double O blocks are used so that realignment can be accomplished without having to perform a tool change if the correct tool is already in the spindle.

Loading the first tool is accomplished by tape sequence O1. The first tool change will automatically cause the Z slide to go to full retract position prior to loading the spindle with tool number T1.

O2   Illustrates the first slide movement at rapid traverse rate (G00) from the random X and Y location where the first tool was loaded. The resultant tool tip path is a **straight line** between the two points in the XY plane, followed by the Z-axis movement. (G00 mode causes X-, Y-, and B-axes movements to occur simultaneously, followed by Z movement.)

This move locates the cutter at the start of the first cut.

Block number two provides the starting coordinates for all axes (X, Y, Z and B), also the mode of operation (G00), spindle speed (S___), spindle (On/off and direction), coolant (On/off and type) (M___) and number of the next tool used (T___). With one exception, any (format) word left out of this block will still contain the last value present in the N/C control memory and the machine will respond according to old data.

Start-up of the control will automatically assume the following functions are in effect:
  G01 Linear Interpolation
  G17 Circular Plane Selection in XY Plane
  G90 Absolute Positioning
  G94 Inches per minute feed rate
The O address permits tape search to locate this block for realignment purposes. For example, if after executing several blocks of tape it should be desired to start over at the beginning, the operator would search block O2 rather than O1 because the first tool is already in the spindle and re-aligning at O1 would cause an undesired tool change.

FIGURE 8-28d
Sample program for first tool change. (Courtesy of Cincinnati Milacron Inc.)

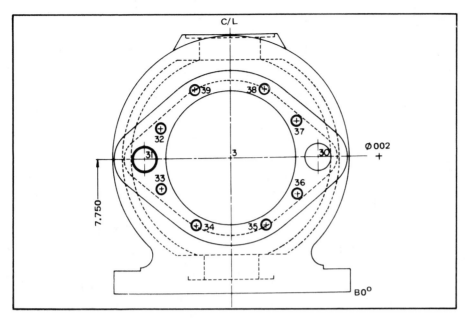

FIGURE 8-28e
Numbered hole positions for 0-side of the sample. (Courtesy of Cincinnati Milacron Inc.)

O/N SEQ.	G PREP. FUNCT.	X± POSITION	Y± POSITION	Z± POSITION	R± POSITION	I/J/K POSITION	A/B/C POSITION P/Q ± WORD	F/E FEED RATE	S SPINDLE SPEED	D/H WORD	T TOOL WORD	M MISC. FUNCT.
N 16		X 0	Y 40000									
N 17	G00			Z 100000								
N 18				Z 36195			B 180000					
N 19	G01	X 50000						F120				

Tape Sequence

N16 Shows slides in their last cutting motion before index.

N17 X- and Y-axes remain at their last cutting position and Z-axis moves at Rapid Traverse rate to a clearance plane which clears any possible interference between cutter and fixture or workpiece.

N18 B-axis rotates to 180 degrees then Z-axis rapids to depth to start milling.

N19 X-axis starts to feed at programmed feed rate.

FIGURE 8-28f
Sample program for index of rotary table. (Courtesy of Cincinnati Milacron Inc.)

O/N SEQ.	G PREP. FUNCT.	X± POSITION	Y± POSITION	Z± POSITION	R± POSITION	I/J/K POSITION	A/B/C P/Q	POSITION ±WORD	F/E FEED RATE	S SPINDLE SPEED	D/H WORD	T TOOL WORD	M MISC. FUNCT.
O182	G00											T16	M 6
O183	G81	X 54650	Y 67949	Z-12500	R 36195		B 180000		F50	S1100		T17	M 3
N184		X 31467	Y 32447										
N185		X- 9551	Y 23335										
N186		X-50553	Y 45953										
N187		X-54165	Y 87051										
N188		X-35147	Y 122553										
N189		X 9551	Y 131665										
N190		X 45053	Y 109047										
N191	G00			Z36295									
O192	G00											T17	M 6
O193	G84	X 45053	Y 109047	Z- 9375	R 36195		B180000		F254	S 440		T19	M 3
N194		X 54650	Y 67949										
N195		X 31467	Y 32447										
N196		X- 9551	Y 23335										
N197		X-50553	Y 45953										
N198		X-54165	Y 87051										
N199		X-35147	Y 122553										
N200		X 9551	Y -131665										
N201	G00			Z36295									
O002	G00											T19	M 6

FIGURE 8-28g

Sample program for drilling and tapping an eight-hole pattern. (Courtesy of Cincinnati Milacron Inc.)

Tape Sequence:

O182 Loads tool (T16) 5/16 x 7/16 subland drill and cosink in the spindle.

O183 G81 (Drill Cycle) causes X-, Y- and B-axes to position at hole 12, first hole of the 8 hole pattern. At the same time the spindle also changes speeds to 1100 rpm in the CW direction (M03). Then Z-axis moves at rapid traverse to the R plane position, feeds at 5.0 ipm (F50) to a depth of 1.2500 inches (Z-12500), and then rapid retracts to the R plane ready to start next operation.

N184 Drills the remaining 7 holes. (13 thru 19). These are the same as the
thru   first hole, except for the X, Y locations, so only the X, Y values are
190   programmed in these blocks.

N191 Retracts the Z-axis to gage height.

O192 Loads (T17) 3/8 - 16 NC tap in the spindle and returns tool (T16) to its proper location in the matrix.

O193 G84 (Tap Cycle) taps hole P19 first of the 8 hole pattern. The slides do not move since they were already in proper position for this hole. The spindle will change speeds to 440 rpm in the CW direction (M03). Then the Z-axis moves at rapid traverse to the R plane position, and feeds at 25.4 ipm (F254) to a depth of 0.9375 inches (Z-9375). At depth the spindle reverses and the Z-axis feeds back to the R plane ready to start next operation.

N194 Taps the remaining 7 holes. These are the same as the first tapped
thru   hole, except for the X, Y location, so only the new X, Y values are
200   programmed in these blocks.

N201 Retracts the Z-axis to gage height.

O202 Tool (T19) is loaded into the spindle for the next operation and T18 returned to storage.

FIGURE 8-28g
Continued

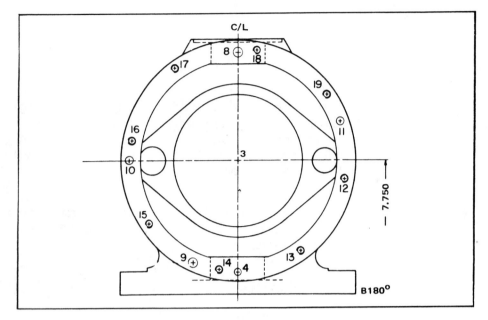

FIGURE 8-28h
Sample workpiece. (Courtesy of Cincinnati Milacron Inc.)

O/N SEQ.	G PREP FUNCT	X± POSITION	Y± POSITION	Z± POSITION	R± POSITION	I/J/K POSITION		A/B/C P/Q POSITION ±WORD		F/E FEED RATE	S SPINDLE SPEED	D/H WORD	T TOOL WORD	M MISC FUNCT.
Ø216	G00												T 2	M 6
Ø217	G00	X 0	Y 77500	Z 56925				B 0			S 555		T 3	M 3
N218	G01			Z 49000						F500				
N219	G02	X-11050	Y 66450			I-11050	J 77500	P 0	Q 10000	F250				
N220		X-22100	Y 77500					P 10000	Q 0	F 76				
N221		X 0	Y 99600			I 0		P 0	Q-10000					
N222		X 22100	Y 77500					P-10000	Q 0					
N223		X 0	Y 55400					P 0	Q 10000					
N224		X-22100	Y 77500					P 10000	Q 0					
N225		X-11050	Y 88550			I-11050		P 0	Q-10000	F250				
N226		X 0	Y 77500							F500				
N227	G01			Z 39425										

FIGURE 8-28i
Sample program for circular milling with CDC. (Courtesy of Cincinnati Milacron Inc.)

A 2-inch diameter end mill is used to circle mill a cored hole to a rough bore dimension of 6.4200", thus eliminating the need to have a boring bar set to this dimension. Since the 2-inch diameter end mill is used elsewhere in the program this technique saves a space in the tool matrix. This technique is valuable when stock removal is heavy or irregular, but especially when the tool matrix is full and no more tools can be added.

The cutter approach path is a semi-circle tangent to the 6.42-inch diameter circle rather than a straight line. This brings the cutter gradually into contact and eliminates "wrap around" which could set up chatter because of the large arc of contact.

Tape Sequence:

O216   Loads 2-inch end mill (T02).

O217   Positions the cutter at the center of the cored hole with CDC off, and rapids Z-axis to gage height. Selects the proper spindle speed and starts the spindle in the CW direction.

N218   Z-axis feeds to depth at 50 ipm.

N219,   G02 produces a clockwise semicircular approach path tangent to the
220   6.50 diameter hole. I and J coordinates define which of the two circles is being used. P and Q coordinates cause the system to offset the cutter from the program path by one-half the CDC value stored in the control memory for this particular tool.

N221,   Each block produces a 90 degree arc of the circle at a feed rate of
2,3,4   7.6 ipm. The feed rate at the periphery is 12 ipm. The new I and J values cause the cutter to follow a new circular path after reaching the tangency point. P and Q values perpetuate the CDC offset already in effect. The vector values are defined at the end of each circular span — intermediate values are calculated by the control.

N225   I and J are programmed here to cause the cut path to follow a semi-circular exit path away from the work. Feed rate is increased since no more metal is being removed. CDC remains "ON" until the cutter is away from the work, to eliminate any marks on the work which might result from cutter deflection.

N226   No P and Q values are present so CDC offset is reduced to zero and the cutter feed rate is stepped up to 50 ipm. Final position is at the center of the cored hole at span end.

N227   Z-axis moves to a new setting for the next cut.

FIGURE 2-28i
Continued

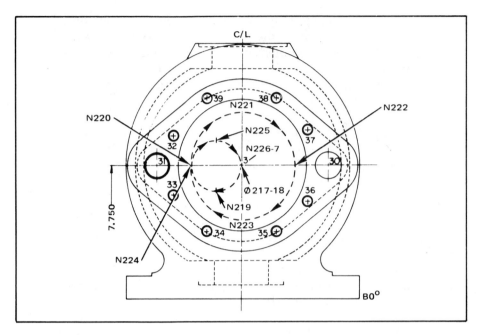

FIGURE 8-28j
Sample part—milling. (Courtesy of Cincinnati Milacron Inc.)

PART NAME			PART NO.		DRAWING NO.	
HOUSING – PUMP VARIABLE DELIVERY			306575			
MACHINE AND CONTROL			REVISION	PROG. BY	DATE	
CIM-X CHANGER 720 with CNC CONTROL					5/8/84	
SETUP AND TOOL INFORMATION			CHECK BY		PAGE OF PAGES	
FIXTURE: USE 2CB-6 8M-434163 SETUP: USE 0.100 IN. GAGE					1	1
TO SET TLC TRAM POSITION X = 0.0000, Y = 3.7500, Z = 6.1000, B = 0						

OPERATION or STATION NUMBER	TOOL DESCRIPTION	ASSEMBLY No.	SET DIM.	REMARKS	CUT SPEED	R.P.M.	FEED	CPT or FD/REV
1	3" CARBIDE END MILL	10011	3.5		300	382	23.0	.052
2	2" CARBIDE END MILL	11030	8.00		224	466	20	.043
3	17/32 DRILL	01022	5.625		80	575	4.9	.008
4	.520 X .5315 BORESIZE DRILL (CARB.)	99015	6.94		250	1800	30	.0167
5	1-1/8 X 1-1/4 SUBLAND DRILL	02014	7.9		82	250	3.0	.012
6	59/64 DRILL	01025	7.4		60	250	3.0	.012
7	3/4 – 14 NPTF PIPE TAP	08006	8.0		40	145	9.4	.065
8	7/16 X 9/16 SUBLAND DRILL	02011	8.0		80	700	5.0	.007
9	1/4 – 18 NPTF PIPE TAP	08009	6.19		40	230	12.3	.056
10	45/64 X 7/8 SUBLAND DRILL	02010	6.9		70	380	4.0	.011
11	1/2 – 14 NPTF PIPE TAP	08005	6.19		40	182	12.5	.069
12	R (.339) X 7/16 SUBLAND DRILL	02009	5.25		80	890	5.4	.006
13	1/8 – 27 NPTF PIPE TAP	08004	4.75		40	377	14.0	.037
14	2.740 BORE AND CHAMFER BAR	09042	6.90		300	416	2.2	.005
15	2.750 BORING BAR	09017	6.75		300	416	2.4	.006
16	5/16 X 7/16 SUBLAND DRILL	02013	5.25		80	977	5.0	.005
17	3/8 – 16 NC TAP	07012	5.56		40	406	25.4	.0625
18	1-7/16 DRILL	01024	8.41		60	158	1.1	.007
19	7/16 DRILL	01020	10.5		60	525	3.1	.006
20	11/32 DRILL	01021	4.88		80	878	5.3	.006
21	27/64 X 9/16 SUBLAND DRILL	02015	5.75		80	725	5.8	.008
22	1/2 – 13 NC TAP	07013	5.94		40	305	23.5	.077
23	1.1725 BORING BAR	09015	6.25		247	805	6.0	.0075
24	6.5000 BORING BAR	09016	3.375		303	176	1.1	.006

FIGURE 8-28k
Sample tooling form.

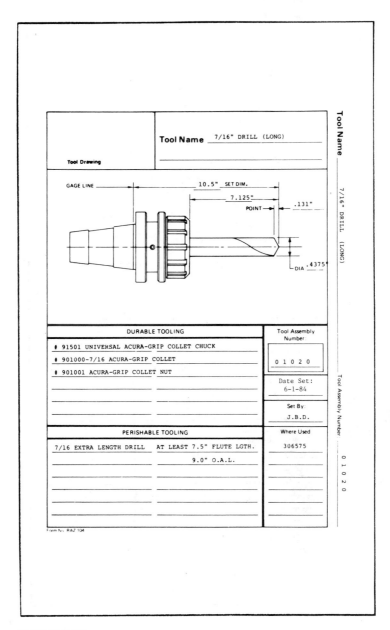

FIGURE 8-28/
Tool assembly drawing.

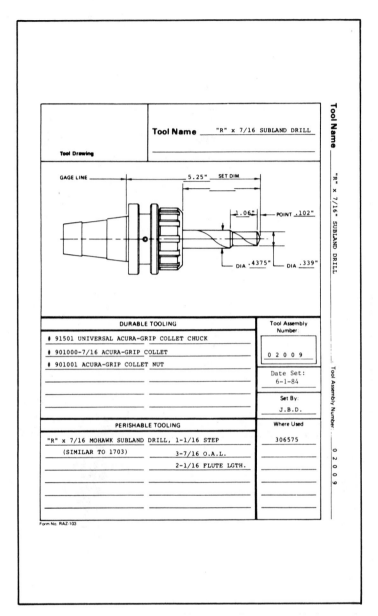

FIGURE 8-28m
Tool assembly drawing.

REVIEW QUESTIONS _____

1. What are the three principal developments that led to the acceptance and use of horizontal machining centers?

2. How are variations in tool lengths handled on a machining center?

3. What types of safety precautions must be considered prior to any tool change and/or table index?

4. List the elements and primary advantages of a pallet system.

5. What are three part-mounting schemes for horizontal machining centers?

6. Explain the difference between tool-dominant and part-dominant programming.

7. Name some optional features available on machining centers.

8. Name the three parts of the surface-sensing probe, and describe some applications.

9. What is the purpose of automatic work changers?

10. What is ship set manufacturing?

11. Explain the function and use of fixture offsets.

CHAPTER 9

Language- and Graphics-Based Programming

OBJECTIVES

After studying this chapter, you will be able to:

- Explain the general structure, format, and capabilities of the APT language
- Describe a postprocessor and identify its major functions
- List some key developments in CAD/CAM since their initial introduction
- Describe the advantages of graphics-based programming systems and explain why they are replacing language-based systems
- List some new CNC features
- Explain shop floor programming (SFP)

PROCESSOR LANGUAGES

Use of the computer to prepare programs for numerically controlled machines began in the early 1960s with language- or processor-based systems. N/C was developed as an answer to some complex aerospace machining problems, such as the aerodynamic curves of blade and airfoil surfaces. Language-based computer programming systems emerged the same way: through a need to machine complex surfaces with a simplified English-like processor language.

At that time, each aerospace company tried to write its own processor language. However, the companies found that the job required more time and manpower than could be committed. Finally, the members of the AIA pooled their resources in a cooperative development project. In 1961, they decided to broaden their scope. They turned over further development to the Illinois Institute of Technology (IIT) for research in Chicago. The APT (Automatic Programmed Tool) system soon emerged, representing over 100 years of development and testing. APT will be discussed in more detail later in this chapter.

Manufacturers soon found that by using a computer and a language-based processor to prepare N/C programs, they could greatly reduce the time and cost of tape preparation, particularly if the part was complex. They also found that processor-based languages produced accurate programs more often than manual programming; in addition, these languages provided error diagnostics relating to format, spelling, and typographical errors at the computer level (where it is generally cheaper to correct) rather than having errors detected at the machine tool itself. This reduced the amount of machine time that might normally be wasted because of tape or programming errors. Figure 9-1 illustrates

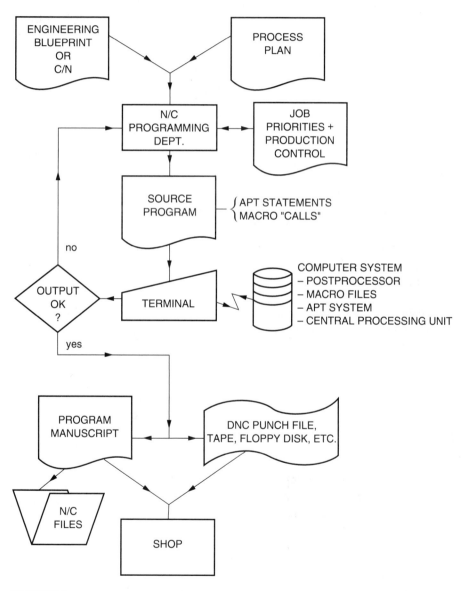

FIGURE 9-1
Flowchart of the development of a language-based N/C program.

the flow of an N/C program when a computer with a language-based processor is being used.

Over time, new and different languages began to be developed, including multiple versions of APT itself. Many of the newly developed languages were more simplistic because the type of parts to be produced were simpler and did not require the full power and capability of APT. Other factors considered when a processor language was selected included types of machine tools used, lot sizes, costs, materials used, and computer type and availability. Although some of these languages are still being used to some degree, many came into prominence and subsequently became obsolete. The list includes the following:

- APT
- AD-APT (Adaptation of APT)
- COMPACT II
- AUTOMAP (Automated Machining Program)
- UNIAPT
- NUFORM
- SPLIT (Sundstrand Processing Language Internally Translated)

APT GENERAL PROCESSOR

Currently, there are several different versions of APT in use throughout industry. Each has its own set of changes or additions to the original APT system. These differences are relatively minor and do not affect the basic parts of the language.

Basically, the APT system is divided into four sections, plus Section 0. Section 0 may be considered the supervisor, or operating system, that controls the flow of information in the various sections of the system. The processing structure is illustrated in figure 9-2.

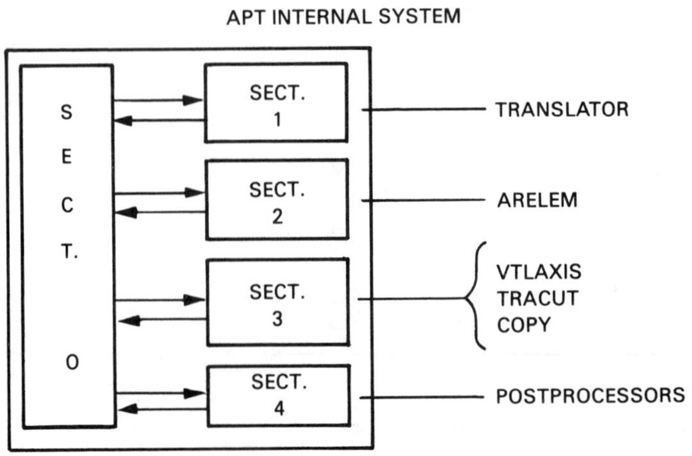

FIGURE 9-2
Sections of APT programming.

In Section 1, the *input translator* phase reads the source statements one at a time. As each statement is processed, it is checked for errors in punctuation, ordering, and syntax as well as for incomplete statements. Any errors detected in this phase cause an *error message* to be sent to the part programmer. The message indicates the type of error and the correction procedure to be followed. The source statements are then separated and classified by type of operation. The necessary data is extracted, rearranged, and passed to the next phase. Certain data dealing entirely with machine tool functions (for example, coolant control, which is not required to compute the cutter center point or path) is passed through to the next phase of the processor.

In Section 2, the *arithmetic* phase of the processor receives the data from the input translator phase. Using a built-in library of subroutines, tables, and symbols, this phase generates the equations that describe a given machining problem. For example, the problem in question might be the intersection of two lines or arcs, the point of tangency of a line and an arc, or the points that describe a circle to be segmented into straight-line cuts within the required tolerance. These equations are solved to find the coordinate values describing the cutting tool's center point in three-dimensional space. These values are formatted into generalized machining instruction sets as the final output of the processor program.

If no errors are found in Section 2, control is passed to Section 3, the *edit* phase. There are three major functions of Section 3. One function, called VTLAXIS, controls the vertical tool axis (a variable tool axis); VTLAXIS is a multiaxis control function that deals with the orientation of the spindle from the vertical position. The other functions, TRACUT and COPY, are used to transform and manipulate the output data of Section 2. If no errors are found in Section 3, or if there is nothing to be done in this section, control is passed on to Section 4.

APT is a multipass processor. It completely processes the input, treating some aspects exclusively (such as spelling and punctuation) before another aspect, such as the calculation of cutter positions, is processed.

Section 4 is the *postprocessor*. The proper postprocessor is selected for the type of machine specified. The data is converted into the proper format by means of a postprocessor program for the specific machine tool which is called out.

POSTPROCESSORS

Postprocessor is the most misunderstood term in numerical control. It has been mistakenly considered a piece of hardware or a separate "black box" sitting off in the corner waiting to "postprocess" some information.

A postprocessor is a set of computer instructions that transforms tool centerline data into machine motion commands using the proper tape code and format required by a specific machine control system. It also performs feed rate calculations, spindle speeds, and auxiliary function commands.

Essentially, each different combination of machine tool and control unit requires its own postprocessor. Whether a language-based processor or a graphics-based system is used to generate the part program, a postprocessor program must still be written for each different machine tool/control unit combination that will be used. Some machine tool and control unit builders contract with outside manufacturing software firms to develop postprocessors for the builder's equipment. Other manufacturers have their own in-house staff for postprocessor development and maintenance.

Because APT and other language- and graphics-based N/C systems are universal, they cannot convert any calculated data into specific tape formats for the machine tool/control unit. The output from the language- or graphics-based system will be the CL, or cutter location, data. This tells where the centerline of the cutter path is located, with respect to the part configuration, within the machine coordinate system. An additional step—postprocessing—is required to adapt the CL output to the particular machine tool/control unit combination that will be used to machine the workpiece. It is the postprocessor output—in the form of the program tape, floppy disk data, or DNC punch file—that will be used on the machine tool. This important relationship is illustrated in figure 9-3.

The primary functions of a postprocessor are as follows:

- To convert cutter location centerline data to the machine coordinate system
- To ensure that the physical limits of the machine are not exceeded (for example, range, feed rate)
- To contain the part to a given tolerance by controlling the amount of overshoot

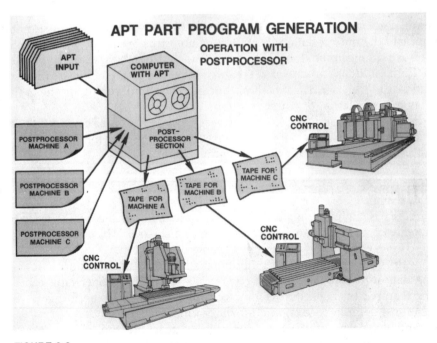

FIGURE 9-3
Relationship of APT program generation, postprocessor, N/C tape, and machine tool.
(Courtesy of Cincinnati Milacron Inc.)

- To output preparatory and miscellaneous functions
- To calculate cutter compensation information
- To generate circular or parabolic points
- To generate error diagnostics when necessary

There are many varieties of postprocessors available. Most are fairly complex and are typically written in FORTRAN. Regardless of what type of postprocessor is used or how it is developed, part programming personnel should thoroughly study the postprocessor documentation in order to become familiar with the capabilities and requirements of the postprocessor. Failure to understand adequately the postprocessor input and output can potentially cause serious damage to the machine tool, cutting tool, and holder; it can also lead to scrapped parts, or even injury to the operator.

Today, postprocessor preparation has been made easier through the development of postprocessor generators. These sophisticated, menu-driven programs allow fast, easy, and userfriendly creation of a postprocessor by having the user answer a series of structured questions about the machine tool and control specifications. Once all the questions have been answered and all special machine tool considerations have been accommodated, the postprocessor is completed and ready to test. Postprocessor generators have become as commonplace today as the postprocessor itself.

A relatively new concept has emerged that virtually eliminates standard postprocessors. Nurtured in the aerospace industry, this feature is called *BCL*, which stands for the use of *binary cutter location data* and is the Electronic Industries Association's published standard RS-494. The importance of BCL is just beginning to be understood and accepted throughout the industry because of the long-term advantages to N/C users and manufacturers.

The purpose of BCL is to standardize formats for exchanging part program data between N/C programming or CAD/CAM systems and CNC controls. Postprocessing, previously accomplished through a mainframe or personal computer, is now moved to a particular machine tool controller. This makes postprocessing a function of the CNC unit through the BCL program (see figure 9-4) and eliminates conventional postprocessing through the computer.

Prior to BCL, the lack of N/C program compatibility among machine tools and controls presented serious problems. These problems centered around the fact that each machine tool had its own particular characteristics—travel ranges, interference zones, maximum feed rates, tool storage capacities, and so forth. Often a program postprocessed for one machine would lack compatibility with another, thereby requiring reprogramming and reprocessing, which uses extra processing time and requires the maintenance of numerous postprocessors serving all machine and control makes and models. BCL, because it is part oriented rather than machine oriented, permits greater flexibility and control of N/C operations.

The following are the primary advantages of BCL:

- Elimination of unique postprocessors. All machines using BCL are programmed the same way, so programmers do not lose efficiency by

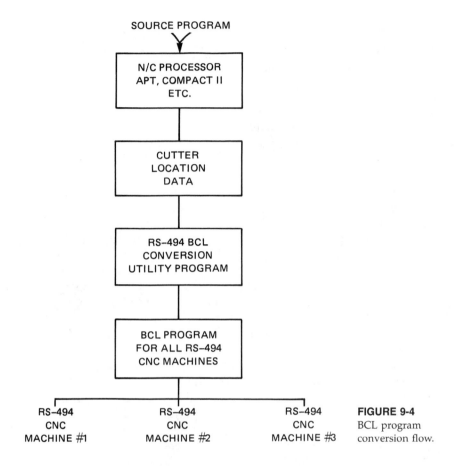

FIGURE 9-4 BCL program conversion flow.

regularly having to relearn differences between different postprocessors for different CNC machines. (Furthermore, the maintenance of a wide array of postprocessors is eliminated.)

• Portability of different part programs among different CNC machines. This refers to the ability to take part programs from one machine to another without any reprocessing or editing. Machine loading and scheduling can therefore be more flexible and responsive. (This leads to a higher CNC machine utilization rate and reduction of work-in-process and programming backlog.)

WRITING AN APT PROGRAM

APT part programming involves three major elements:

1. Definition and symbolic naming of geometric points and surfaces representing part size and configuration.

2. Specification of cutting tool and action or tool motion statements. (These statements move the cutter to the points or along the defined geometric surfaces.)

3. Specification of conditions required at the machine tool, such as spindle speeds, feed rates, and other auxiliary function commands. (These correspond to the independent postprocessor being used.)

The APT geometry statement primarily consists of three parts (symbol, geometric element, and description) in the following format:

$$Symbol = Geometric\ element/Description$$

Consider an example:

$$C1 = Circle/center,\ P2,\ radius,\ R$$

The first part is a symbol, which is an arbitrary name assigned to a particular geometric element. This symbol is then equated to the definition (the second part of the statement), which is a major word such as point, line, circle, etc. The major word defines the type of surface or geometric element that the symbol represents. The third part of the APT statement is the actual description, which consists of minor words or modifiers and numerical values of the point, line, circle, etc. These words position the element in space and determine its specific size.

Motion statements in the APT language are typical of statements that might be used in directing a person to walk around the block or through town (go left, go right, left side, right side). English-like APT vocabulary words such as TLLFT (tool left), TLRGT (tool right), GOLFT (go left), and GORGT (go right) are used to direct the tool path relative to the defined geometry. As discussed earlier, Section 1 of the APT processor checks for spelling, punctuation, and syntax errors. Section 2 makes the mathematical calculations and generates the CL data, and Section 3 checks for any tool axis orientation, copying, or transformation of the CL data. Once these are completed, the CL data is then ready to be passed to the designated postprocessor for tape data generation.

Here is an example of an APT tool motion statement:

Positional modifier	Directional modifier	Drive surface	Modifier	Check surface
TLLFT	GOLFT	/L1	TANTO	C1

When programming using a language-based processor such as APT, the part programmer must first geometrically define the part to be produced and its various elements and surfaces, selecting the best available format from the APT language. There are numerous formats available, and each must be used in exactly the same way it is provided in the APT vocabulary. In providing the geometric definitions of the part, tool motion statements, and ancillary commands, the APT part programmer communicates to the computer using the specific APT vocabulary. In this vocabulary, there are approximately 260 words, including punctuation. The part programmer does not have the freedom to invent new words or modify any APT definition statement.

Two simple APT programs showing both APT definition and tool motion statements are given in figures 9-5 and 9-6.

A much more detailed, typical, and complete APT program, along with its postprocessor printout, is shown in figure 9-7. Included in this figure are the operation sheets and commentary contained in the program.

There are many reasons why APT continues to be the mainstay of language-based processors. Principal among these are the following:

- APT has a wide breadth of application use and is still the most advanced multiaxis processor language. The various versions currently in use throughout industry are mature, reliable processors capable of producing predictable, dependable output results for machine tools.

- The voluminous number of part programs already programmed in APT makes conversion to graphics-based programming for parts already programmed in APT a non–value-added effort.

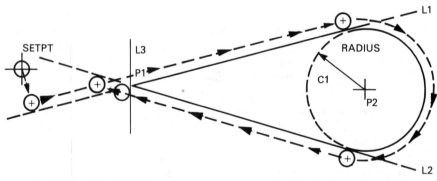

APT GEOMETRY STATEMENTS
SETPT = POINT/X, Y, Z
  P1 = POINT/X, Y, Z
  P2 = POINT/X, Y, Z
  C1 = CIRCLE/CENTER, P2, RADIUS, R
  L1 = LINE/P1, LEFT, TANTO, C1
  L2 = LINE/P1, RIGHT, TANTO, C1
  L3 = LINE/P1, ATANGL, 90
CUTTER/.5

APT MOTION STATEMENTS
FROM/ SETPT
RAPID
GO/ TO, L1
FEDRAT/20
TLLFT, GOLFT/L1, TANTO, C1
GOFWD/C1, TANTO, L2
GOFWD/L2, PAST, L3
GOTO/SETPT
FINI

FIGURE 9-5
Simple APT program.

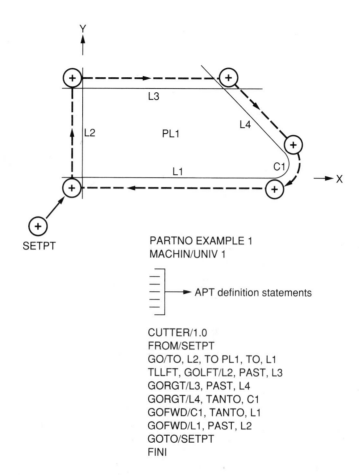

PARTNO EXAMPLE 1
MACHIN/UNIV 1

APT definition statements

CUTTER/1.0
FROM/SETPT
GO/TO, L2, TO PL1, TO, L1
TLLFT, GOLFT/L2, PAST, L3
GORGT/L3, PAST, L4
GORGT/L4, TANTO, C1
GOFWD/C1, TANTO, L1
GOFWD/L1, PAST, L2
GOTO/SETPT
FINI

FIGURE 9-6
A simple APT program
illustrating more tool
motion statements.

- APT offers a simple means of handling family-of-parts situations through the use of extensive macros or subroutines (programs within a program). Macros have been set up where only dimensional variables need to be changed so as to program an entirely new part. These macros are easily handled in APT and other processor languages, but are much more difficult for graphics-based systems.

Although APT and some other language-based processors will continue to be used in the future for the reasons just discussed, they are slowly being replaced by the more productive, interactive, and user-friendly graphics-based systems.

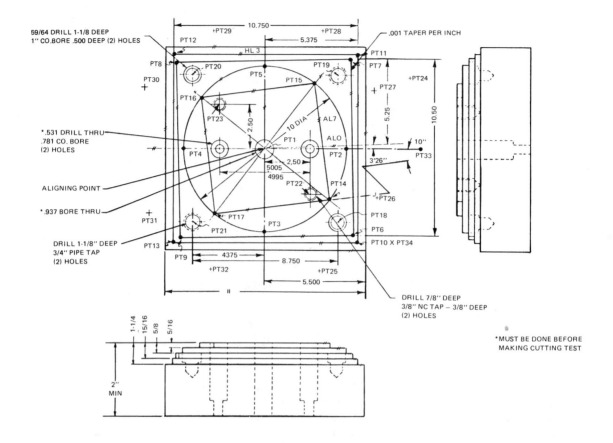

SEQUENCE OF OPERATIONS
1. Align Point — 1" bored hole (X000000 Y000000) P1
2. Mill Off Corners of Part — 5/8 deep, 2 passes, 5/16 stock removal each pass 2" carb. end mill, 8 ipm. P25–P32
3. Mill Circle — 5/8 deep, 2 passes, 5/16 stock removal each pass, 2" carb. end mill, leave .020 on radius for finish cut, 12 ipm. P2–P5
4. Mill Square Inside Circle — 5/16 deep, 1 pass, 2" carb. end mill, square at 10° angle, leave .020 on sides for Finish cut, 8 ipm. P14–P17
5. Mill (Taper) Square — 5/16 depth of cut (15/16 deep from top of part) 1 pass, 2" carb. end mill, leave .020 on sides for Finish cut, 14 ipm. P6–P9
6. Mill Square — 5/16 depth of cut (1-1/4 deep from top of part) 1 pass, 2" carb. end mill, leave .020 on sides for Finish cut, 14 ipm. P10–P13
7. Mill Square To Finish Dimension — 2" carb. end mill, 1 pass 20 ipm. P10–P13
8. Mill Square With Taper To Finish Dimension — 2" carb. end mill, 1 pass, 20 ipm. P6–P9
9. Mill Circle to Finish Dimension — 2" carb. end mill, 1 pass, 16 ipm. P2–P5

FIGURE 9-7
Detailed APT program. (Courtesy of Cincinnati Milacron Inc.)

10. Mill Square Inside Circle to Finish Dimension — 2″ carb. end mill, 1 pass, 16 ipm. P14–P17
11. Drill 4 Holes — 1-1/8″ deep, 59/64 dia. drill. P18–P21
12. Pipe Tap 2 Holes — 3/4 — 14 pipe tap. P19 and P21
13. Bore 2 Holes — 1/2″ deep, 1″ dia. bore. P18 and P20
14. Drill 2 Holes — 7/8″ deep, 5/16 dia. drill. P22 and P23
15. Tap 2 Holes — 3/8″ deep, 3/8 — 16 n.c. tap. P22 and P23
16. Unload.

RAE-103

```
        N/C 360 APT VERSION 4, MODIFICATION 2     DATE= 71.284 TIME OF DAY IN HRS./MIN./SEC IS 12/32/46.20


  THE MACRO    CIR   USES     66  LOCATIONS IN CANON
  THE MACRO   CIRSQR USES     90  LOCATIONS IN CANON
  THE MACRO   ANGSQR USES     96  LOCATIONS IN CANON
  THE MACRO    SQUR  USES     96  LOCATIONS IN CANON

                            TABLE USAGE DURING INPUT TRANSLATION


                      PASS ONE                              PASS TWO
                 ALLOCATED      USED                   DYNAMIC ALLOCATION
          VST     2750          124              VST                125
          PTPP    2225          272              PTPP               272
          CANON   2225          348              SCALARS             11
                                                 CANON             5792



    1  PARTNO   CPG COMPOSITE TEST PART FOR THE CIM-X 720 WITH TOOL COMP        CPGC0000
    2  $$                                                                       CPGC0010
    3           MACHIN/ CINAC1, 156000, CIRCUL, 10, 12.2                        CPGC0020
    4  $$                                                                       CPGC0030
    5           ORIGIN/ 0, 0, 3           $$ ORIGIN FOR CONTROL WITH TOOL COMP  CPGC0040
    6  $$                                                                       CPGC0050
    7  PT1    = POINT/0,0,0                                                     CPGC0060
    8  ULOAD  = POINT/15, 6,0                                                   CPGC0070
    9  C1     = CIRCLE/0,0,5                                                    CPGC0080
   10  HLO    = LINE/PT1,ATANGL,0.0                                            CPGC0090
   11  VLO    = LINE/PT1,ATANGL,90.0                                           CPGC0100
   12  ALO    = LINE/PT1,ATANGL,.0572222                                       CPGC0110
   13  AL1    = LINE/PARLEL,ALO,YLARGE,5.25                                    CPGC0120
   14  AL2    = LINE/PARLEL,ALO,YSMALL,5.25                                    CPGC0130
   15  AL3    = LINE/PARLEL,(LINE/PT1,PERPTO,ALO),XLARGE,5.25                  CPGC0140
   16  AL4    = LINE/PARLEL,AL3,XSMALL,10.50                                   CPGC0150
   17  AL5    = LINE/(POINT/YLARGE, INTOF,(ALA=LINE/PT1,ATANGL,55),C1),ATANGL,10 CPGC0160
   18  AL6    = LINE/(POINT/YSMALL, INTOF,ALA,C1),ATANGL, 10                   CPGC0170
   19  ALB    = LINE/PT1,PERPTO,ALA                                            CPGC0180
   20  AL7    = LINE/(POINT/XLARGE, INTOF,ALB,C1),PERPTO,AL5                   CPGC0190
   21  AL8    = LINE/(POINT/XSMALL, INTOF,ALB,C1),PERPTO,AL5                   CPGC0200
   22  VL1    = LINE/PARLEL,VLO,XLARGE,5.375                                   CPGC0210
   23  VL2    = LINE/PARLEL,VLO,XSMALL,5.375                                   CPGC0220
   24  VL3    = LINE/PARLEL,VLO,XLARGE,5.5                                     CPGC0230
   25  VL4    = LINE/PARLEL,VLO,XSMALL,5.5                                     CPGC0240
   26  HL1    = LINE/PARLEL,HLO,YSMALL,5.375                                   CPGC0250
   27  HL2    = LINE/PARLEL,HLO,YSMALL,5.5                                     CPGC0260
   28  HL3    = LINE/PARLEL,HLO,YLARGE,5.375                                   CPGC0270
   29  HL4    = LINE/PARLEL,HLO,YLARGE,5.5                                     CPGC0280
   30  PT18   = POINT/4.375,-4.375,-.625                                       CPGC0290
   31  PT19   = POINT/4.375, 4.375,-.625                                       CPGC0300
   32  PT20   = POINT/-4.375,4.375,-.625                                       CPGC0310
   33  PT21   = POINT/-4.375,-4.375,-.625                                      CPGC0320
   34  PT22   = POINT/2.5,-2.5,0.0                                             CPGC0330
   35  PT23   = POINT/-2.5,2.5,0.0                                             CPGC0340
   36  PT25   = POINT/3.1358,-7.0,-.3125                                       CPGC0350
   37  PT26   = POINT/6.7600,-3.1358,-.3125                                    CPGC0360
   38  PT27   = POINT/6.7600,3.1358,-.3125                                     CPGC0370
   39  PT28   = POINT/3.1358,7.0,-.3125                                        CPGC0380
   40  PT29   = POINT/-3.1358,7.0,-.3125                                       CPGC0390
   41  PT30   = POINT/-6.7600,3.1358,-.3125                                    CPGC0400
   42  PT31   = POINT/-6.7600,-3.1358,-.3125                                   CPGC0410
```

VIII-D 9

FIGURE 9-7

Continued...

```
43    PT32    = POINT/-3.1358,-7.0,-.3125                                    CPGC0420
44    PT33    = POINT/3.1358,-7.0,0.0                                        CPGC0430
45    C1      = LINE/PT1,PT25                                                CPGC0440
46    PL0     = PLANE/0,0,1,0                                                CPGC0450
47    PL1     = PLANE/0,0,1,-.3125                                           CPGC0460
48    PL2     = PLANE/0,0,1,-.625                                            CPGC0470
49    PL3     = PLANE/0,0,1,-.9375                                           CPGC0480
50    PL4     = PLANE/0,0,1,-1.25                                            CPGC0490
51    $$                                                                     CPGC0500
52    $$ THE FOLLOWING ARE SYMBOLIC FEEDRATES--CHECK TO SEE IF THEY SHOULD   CPGC0510
53    $$ BE CHANGED PRIOR TO PROCESSING.                                     CPGC0520
54    $$                                                                     CPGC0530
55    RFED1   = 8             $$ RGH CORNERS OF PART, INSIDE SQUARE          CPGC0540
56    RFED2   = 12            $$ RGT CIRCLE                                  CPGC0550
57    RFED3   = 14            $$ RGT OUTSIDE SQUARE                          CPGC0560
58    FFED1   = 20            $$ FINISH MILL SQUARE                          CPGC0570
59    FFED2   = 16            $$ FINSIH MILL CIRCLE                          CPGC0580
60    FFED3   = 20            $$ FINISH TAPERD SQUARE                        CPGC0590
61    DRFED   = 15            $$ DRILL                                       CPGC0600
62    TPFED1  = 15            $$ TAPERD PIPE TAP                             CPGC0610
63    TPFED2  = 21            $$ N.C.TAP                                     CPGC0620
64    BRFED   = 10            $$ BORE                                        CPGC0630
65    RAPD    = 150           $$ RAPID                                       CPGC0640
66    $$                                                                     CPGC0650
67    $$         MACROS                                                      CPGC0660
68    $$                                                                     CPGC0670
69    CTR     = MACRO/DIA,PLN,FED                                            CPGC0680
70            CUTTER/DIA                                                     CPGC0690
71            FEDRAT/FED                                                     CPGC0700
72            SPINDL/880,CLW                                                 CPGC0710
73            PSIS/PLN                                                       CPGC0720
74            FROM/PT33                                                      CPGC0730
75            INDIRP/PT1                                                     CPGC0740
76            GO/TO,C1                                                       CPGC0750
77            TLRGT,GORGT/C1,ON,2,INTOF,(LINE/PT1,PTZ5)                      CPGC0760
78            TERMAC                                                         CPGC0770
79    $$                                                                     CPGC0780
80    CIRSQR  = MACRO/DIAM,FEDR                                              CPGC0790
81            CUTTER/DIAM                                                    CPGC0800
82            CYCLE/MILLRP,DEEP,0.0                                          CPGC0810
83            SPINDL/880,CLW                                                 CPGC0820
84            FEDRAT/FEDR                                                    CPGC0830
85            ZSURF/PL1                                                      CPGC0840
86            GOTO/(POINT/INTOF,AL8,(LINE/PARLEL,AL7,XLARGE,1.25))           CPGC0850
87            PSIS/PL1                                                       CPGC0860
88            INDIRP/PT1                                                     CPGC0870
89            GO/TO,AL7                                                      CPGC0880
90            TLRGT,GORGT/AL7                                                CPGC0890
91            GOLFT/AL5                                                      CPGC0900
92            GOLFT/AL8                                                      CPGC0910
93            GOLFT/AL6,PAST,AL7                                             CPGC0920
94            TERMAC                                                         CPGC0930
95    $$                                                                     CPGC0940
96    ANGSQR  = MACRO/DIAMT,FDRT                                             CPGC0950
97            CUTTER/DIAMT                                                   CPGC0960
98            SPINDL/880,CLW                                                 CPGC0970
99            FEDRAT/FDRT                                                    CPGC0980
100           CYCLE/MILLRP,DEEP,0.0                                          CPGC0990
101           ZSURF/PL3                                                      CPGC1000
102           GOTO/(POINT/INTOF,(LINE/PARLEL,VL3,XLARGE,1.5),$               CPGC1010
                 (LINE/PARLEL,HL2,YSMALL,1.5))                               CPGC1020
103           PSIS/PL3                                                       CPGC1030
104           INDIRP/PT1                                                     CPGC1040
105           GO/TO,AL3                                                      CPGC1050
106           TLRGT,GORGT/AL3                                                CPGC1060
107           GOLFT/AL1                                                      CPGC1070
108           GOLFT/AL4                                                      CPGC1080
109           GOLFT/AL2,PAST,AL3                                             CPGC1090
110           TERMAC                                                         CPGC1100
111   $$                                                                     CPGC1110
112   SQUR    = MACRO/DTR,FRAT                                               CPGC1120
113           CUTTER/DTR                                                     CPGC1130
114           SPINDL/880,CLW                                                 CPGC1140
115           CYCLE/MILLRP,DEEP,0.0                                          CPGC1150
116           ZSURF/PL4                                                      CPGC1160
117           GOTO/(POINT/INTOF,(LINE/PARLEL,VL3,XLARGE,1.5),$               CPGC1170
                 (LINE/PARLEL,HL2,YSMALL,1.5))                               CPGC1180
118           FEDRAT/FRAT                                                    CPGC1190
119           PSIS/PL4                                                       CPGC1200
120           INDIRP/PT1                                                     CPGC1210
121           GO/TO,VL1                                                      CPGC1220
122           TLRGT,GORGT/VL1                                                CPGC1230
```

FIGURE 9-7

Continued...

```
123        GOLFT/HL3                                                    CPGC1240
124        GOLFT/VL2                                                    CPGC1250
125        GOLFT/HL1,PAST,VL1                                           CPGC1260
126        TERMAC                                                       CPGC1270
127   $$                                                                CPGC1280
128   $$   START OF PROGRAM                                             CPGC1290
129   $$                                                                CPGC1300
130   PPRINT                                                            CPGC1310
131   PPRINT  ALIGN POINT IS AT 1INCH BORED HOLE (X000000 Y000000)      CPGC1320
132   PPRINT                                                            CPGC1330
133   PPRINT  LOAD 2-INCH DIA 4-FLUTE END MILL                          CPGC1340
134   PPRINT                                                            CPGC1350
135   PPRINT  MILL CORNERS OFF PART- 2 PASSES,5/16 DEPTH EACH PASS      CPGC1360
136        LOADTL / 1,1                                                 CPGC1370
137   $$                                                                CPGC1380
138        CLRSRF / XYPLAN, 3                                           CPGC1390
139        ROTABL / ATANGL, 90                                          CPGC1400
140        SPINDL / 880, CLW                                            CPGC1410
141        COOLNT / MIST                                                CPGC1420
142        FEDRAT / RFED1                                               CPGC1430
143        FROM/PT1                                                     CPGC1440
144   CYCLE/MILLRP,DEEP,0.0                                             CPGC1450
145        GOTO/PT25                                                    CPGC1460
146        GOTO/PT26                                                    CPGC1470
147        RAPID                                                        CPGC1480
148        GOTO/PT27                                                    CPGC1490
149        GOTO/PT28                                                    CPGC1500
150        RAPID                                                        CPGC1510
151        GOTO/PT29                                                    CPGC1520
152        GOTO/PT30                                                    CPGC1530
153        RAPID                                                        CPGC1540
154        GOTO/PT31                                                    CPGC1550
155        GOTO/PT32                                                    CPGC1560
156   CYCLE/MILLRP,DEEP,.3125                                           CPGC1570
157        GOTO/PT32                                                    CPGC1580
158        GOTO/PT31                                                    CPGC1590
159        RAPID                                                        CPGC1600
160        GOTO/PT30                                                    CPGC1610
161        GOTO/PT29                                                    CPGC1620
162        RAPID                                                        CPGC1630
163        GOTO/PT28                                                    CPGC1640
164        GOTO/PT27                                                    CPGC1650
165        RAPID                                                        CPGC1660
166        GOTO/PT26                                                    CPGC1670
167        GOTO/PT25                                                    CPGC1680
168   PPRINT                                                            CPGC1690
169   PPRINT ROUGH MILL CIRCLE--2 DEPTH PASSES-- LEAVE .020 EXCESS--RPM 880  CPGC1700
170   PPRINT                                                            CPGC1710
171        CYCLE / MILLFD, DEEP, 0.3125                                 CPGC1720
172        CALL/CIR,DIA=2.04,PLN=PLO,FED=RFED2                          CPGC1730
                                                                        CPGC1740
174        CYCLE/MILLFD,DEEP,.625                                       CPGC1750
175        CALL/CIR,DIA=2.04,PLN=PLO,FED=RFED2                          CPGC1760
176        GOTO/PT33                                                    CPGC1770
177   PPRINT                                                            CPGC1780
178   PPRINT MILL SQUARE INSIDE CIRCLE--ALLOW .020 EXCESS--RPM 880      CPGC1790
179   PPRINT                                                            CPGC1800
180        CALL/CIRSQR,DIAM=2.04,FEDR=RFED1                             CPGC1810
181   PPRINT                                                            CPGC1820
182   PPRINT MILL TAPERD SQUARE--1 PASS-- LEAVE .020 EXCESS--RPM 880    CPGC1830
183   PPRINT                                                            CPGC1840
184        CALL /ANGSQR,DIAMT=2.04,FORT=RFED3                           CPGC1850
185   PPRINT                                                            CPGC1860
186   PPRINT MILL OUTSIDE SQUARE--1 PASS-- LEAVE .020 EXCESS--RPM 880   CPGC1870
187   PPRINT                                                            CPGC1880
188        CALL/SQUR,DTR=2.04,FRAT=RFED3                                CPGC1890
189   PPRINT                                                            CPGC1900
190   PPRINT RETRACT TO BACK LIMIT--POSITION AT UNLOAD POINT            CPGC1910
191   PPRINT INSPECT CUTTER AND PART                                    CPGC1920
192   PPRINT                                                            CPGC1930
193        RAPID                                                        CPGC1940
194        GOTO/ULOAD                                                   CPGC1950
195        OPSTOP                                                       CPGC1960
196   PPRINT                                                            CPGC1970
```

FIGURE 9-7

Continued...

```
197   PPRINT MILL SQUARE--FINISH PASS--RPM 880              CPGC1980
198   PPRINT                                                CPGC1990
199          CALL/SQDR,DTR=2.0,FRAT=FFED3                   CPGC2000
200   PPRINT                                                CPGC2010
201   PPRINT MILL TAPERD SQUARE -- FINISH PASS--RPM 880     CPGC2020
202   $$                                                    CPGC2030
203          CALL/ANGSQR,DIAMT=2.0,FDRT=FFED3               CPGC2040
204   PPRINT                                                CPGC2050
205   PPRINT MILL CIRCLE--FINISH PASS--1 PASS--RPM 880      CPGC2060
206   $$                                                    CPGC2070
207          CYCLE / MILLFD, DEEP, 0.625                    CPGC2080
208              GOTO/PT33,30                               CPGC2090
209          CALL/CIR,DIA=2.0,PLN=PLO,FED=FFEDZ             CPGC2100
210   PPRINT                                                CPGC2110
211   PPRINT MILL SQUARE INSCRIBED IN CIRCLE--FINISH PASS--RPM 880  CPGC2120
212   $$                                                    CPGC2130
213          CALL/CTRSQR,DIAM=2.0,FEDR=FFEDZ                CPGC2140
214   PPRINT                                                CPGC2150
215   PPRINT LOAD TOOL 2 -- 59/64 DRILL                     CPGC2160
216   $$                                                    CPGC2170
217          LOADTL / 2,2                                   CPGC2180
218   $$                                                    CPGC2190
219          SPINDL / 1400, CLW                             CPGC2200
220          COOLNT / FLOOD                                 CPGC2210
221   CYCLE/DRILL,DEEP,1.1250,IPM,18.0                      CPGC2220
222          GOTO/PT18                                      CPGC2230
223          RETRCT                                         CPGC2240
224          GOTO/PT21                                      CPGC2250
225          RETRCT                                         CPGC2260
226          GOTO/PT20                                      CPGC2270
227          RETRCT                                         CPGC2280
228          GOTO/PT19                                      CPGC2290
229   PPRINT                                                CPGC2300
230   PPRINT LOAD TOOL 3 -- 3/4-14 PIPE TAP                 CPGC2310
231   $$                                                    CPGC2320
232          LOADTL / 3,3                                   CPGC2330
233          SELCTL / 4                                     CPGC2340
234          SPINDL / 175, CLW                              CPGC2350
235          CYCLE/TAP  ,DEEP,1.0,IPM,3.0                   CPGC2360
236          GOTO/PT21                                      CPGC2370
237          RETRCT                                         CPGC2380
238          GOTO/PT19                                      CPGC2390

239   PPRINT                                                CPGC2400
240   PPRINT LOAD TOOL 4 -- 1.0 BORE(CARB. TIP)             CPGC2410
241   $$                                                    CPGC2420
242          LOADTL / 4,4                                   CPGC2430
243   $$                                                    CPGC2440
244          SPINDL / 1755, CLW                             CPGC2450
245          CYCLE/BORE,DEEP,0.50,IPM,6.0                   CPGC2460
246          GOTO/PT18                                      CPGC2470
247          RETRCT                                         CPGC2480
248          GOTO/PT20                                      CPGC2490
249   PPRINT                                                CPGC2500
250   PPRINT LOAD TOOL 5 -- 5/16 DRILL                      CPGC2510
251   $$                                                    CPGC2520
252          LOADTL / 5,5                                   CPGC2530
253   $$                                                    CPGC2540
254          SPINDL / 2785, CLW                             CPGC2550
255          CYCLE/DRILL,DEEP,0.8750,IPM,12.0               CPGC2560
256          GOTO/PT22                                      CPGC2570
257          GOTO/PT23                                      CPGC2580
258   PPRINT                                                CPGC2590
259   PPRINT LOAD TOOL 6 -- 3/8-16 N.C. TAP                 CPGC2600
260   $$                                                    CPGC2610
261          LOADTL / 6,6                                   CPGC2620
262          SPINDL / 220, CLW                              CPGC2630
263          CYCLE/TAP,DEEP,0.3750,IPM,3.0                  CPGC2640
264          GOTO/PT22                                      CPGC2650
265          GOTO/PT23                                      CPGC2660
266          RAPID                                          CPGC2670
267          GOTO/ ULOAD                                    CPGC2680
268          END                                            CPGC2690
269          FINI                                           CPGC2700
```

FIGURE 9-7

Continued...

RAE-103

71.284   C I N A C 1   156000 / A P T   I N C H   P O S T P R O C E S S O R   LEVEL A   PAGE   1

A C R A M A T I C   3 3 5 - D

O/N G	X	Y	Z	I	J	F	R	S	B	W	M S	CLNO	RPM	TIME
CPG COMPOSITE TEST PART FOR THE CIM-X 720 WITH TOOL COMP														
LEADER/ 72.0														

ALIGN POINT IS AT 1INCH BORED HOLE (X000000 Y000000)

LOAD 2-INCH DIA 4-FLUTE END MILL

MILL CORNERS OFF PART-- 2 PASSES, 5/16 DEPTH EACH PASS

O/N G	X	Y	Z	I	J	F	R	S	B	W	M S	CLNO	RPM	TIME
$0 1 G80 X&	0 Y&	0 Z	0			F 1 R	0				M06$	32	110	.283
0 2 G80 X&	0 Y&	0 Z	0			F 80 R	30000 S 12 B 90000 W 1				M17$	32	880	.014
0 3 G80 X&	31358 Y-	70000 Z	0			F 80 R	34125 S 12 B 90000 W 1				M17$	36	880	.040
N 4 G79 X&	67600 Y-	31358									$	38	880	.662
N 5 G80	Y&	31358									$	42	880	.031
N 6 G79 X&	31358 Y&	70000									$	44	880	.662
N 7 G80 X-	31358										$	48	880	.031
N 8 G79 X-	67600 Y&	31358									$	50	880	.662
N 9 G80	Y&	31358									$	54	880	.031
N 10 G79 X-	31358 Y-	70000									$	56	880	.662
0 11 G80 X-	31358 Y-	70000 Z	0			F 80 R	37250 S 12 B 90000 W 1				M17$	60	880	.001
N 12 G79 X-	67600 Y-	31358									$	62	880	.662
N 13 G80	Y&	31358									$	66	880	.031
N 14 G79 X-	31358 Y&	70000									$	68	880	.662
N 15 G80 X&	31358										$	72	880	.031
N 16 G79 X&	67600 Y&	31358									$	74	880	.662
N 17 G80	Y-	31358									$	78	880	.031
N 18 G79 X&	31358 Y-	70000									$	80	880	.662

ROUGH MILL CIRCLE--2 DEPTH PASSES-- LEAVE .020 EXCESS--RPM 880

O/N G	X	Y	Z	I	J	F	R	S	B	W	M S	CLNO	RPM	TIME
N 19 G80		Z	3125			F 120 R	30000				$	98	880	.003
0 20 G80 X&	31358 Y-	70000 Z	3125			F 120 R	30000 S 12 B 90000 W 1				M17$	98	880	.000
N 21 G79		Z	3125								$	98	880	.026
N 22	X&	24611 Y-	54939								$	101	880	.163
N 23 G03 X&	60200 Y&	0		I&	0 J&	0					$	107	880	.576
N 24 G03 X-	0 Y&	60200		I&	0 J&	0					$	107	880	.788
N 25 G03 Y-	60200 Y-	0		I&	0 J&	0					$	107	880	.788
N 26 G03 X&	0 Y-	60200		I&	0 J&	0					$	107	880	.788
N 27 G03 X&	24611 Y-	54939		I&	0 J&	0					$	107	880	.211
N 28 G80 X&	31358 Y-	70000					R	33125			$	110	880	.009
N 29 G79		Z	3125								$	110	880	.026
N 30 G80											$	112	880	.000
N 31		Z	6250				R	30000			$	122	880	.001
0 32 G80 X&	31358 Y-	70000 Z	6250			F 120 R	30000 S 12 B 90000 W 1				M17$	122	880	.000
N 33 G79		Z	6250								$	122	880	.052
N 34	X&	24611 Y-	54939								$	125	880	.189
N 35 G03 X&	60200 Y&	0		I&	0 J&	0					$	131	880	.576
N 36 G03 X-	0 Y&	60200		I&	0 J&	0					$	131	880	.788

MACHINING TIME   10.815 MINUTES   TAPE LENGTH   6.70 FEET

RAE-103

71.284   C I N A C 1   156000 / A P T   I N C H   P O S T P R O C E S S O R   LEVEL A   PAGE   2

CPG COMPOSITE TEST PART FOR THE CIM-X 720 WITH TOOL COMP

O/N G	X	Y	Z	I	J	F	R	S	B	W	M S	CLNO	RPM	TIME
N 37 G03 X-	60200 Y-	0		I&	0 J&	0					$	131	880	.788
N 38 G03 X&	0 Y-	60200		I&	0 J&	0					$	131	880	.788
N 39 G03 X&	24611 Y-	54939		I&	0 J&	0					$	131	880	.211
N 40 G79 X&	31358 Y-	70000									$	134	880	.189

MILL SQUARE INSIDE CIRCLE--ALLOW .020 EXCESS--RPM 880

O/N G	X	Y	Z	I	J	F	R	S	B	W	M S	CLNO	RPM	TIME
N 41 G80											$	145	880	.000
0 42 G80 X&	55438 Y-	38818 Z	0			F 80 R	34125 S 12 B 90000 W 1				M17$	151	880	.021
N 43 G79 X&	52774 Y-	36953									$	155	880	.040
N 44	X&	36953 Y&	52774								$	157	880	1.138
N 45	X-	52774 Y&	36953								$	159	880	1.138
N 46	X-	36953 Y-	52774								$	161	880	1.138
N 47	X&	52774 Y-	36953								$	163	880	1.138

MILL TAPERD SQUARE--1 PASS--LEAVE .020 EXCESS--RPM 880

O/N G	X	Y	Z	I	J	F	R	S	B	W	M S	CLNO	RPM	TIME
0 48 G80 X&	70000 Y-	70000 Z	0			F 140 R	40375 S 12 B 90000 W 1				M17$	181	880	.021
N 49 G79 X&	62763 Y-	62763									$	185	880	.073
N 50	X&	62637 Y-	62763								$	187	880	.896
N 51	X-	62763 Y&	62637								$	189	880	.895
N 52	X-	62637 Y-	62763								$	191	880	.895
N 53	X&	62763 Y-	62637								$	193	880	.895

FIGURE 9-7
Continued...

```
MILL OUTSIDE SQUARE--1 PASS-- LEAVE .020 EXCESS--RPM 880
----------------------------------------------------------------------------------------------------
 0 54 G80 X&  70000 Y-  70000 Z      0              F 140 R  43500 S 12 B 90000 W 1 M17$  209   880    .006
 N 55 G79 X&  63950 Y-  63950                                                         $  215   880    .061
 N 56         Y&  63950                                                               $  217   880    .913
 N 57     X-  63950                                                                   $  219   880    .913
 N 58         Y-  63950                                                               $  221   880    .913
 N 59     X&  63950                                                                   $  223   880    .913
----------------------------------------------------------------------------------------------------
RETRACT TO BACK LIMIT--POSITION AT UNLOAD POINT
INSPECT CUTTER AND PART
----------------------------------------------------------------------------------------------------
 N 60 G80                                           R  31000                          $  236   880    .006
 N 61     X& 150000 Y& -60000                                                         $  236   880    .075
 N 62                                                                            M01$  238   880    .000
----------------------------------------------------------------------------------------------------
  MACHINING TIME   24.893 MINUTES              TAPE LENGTH   11.12 FEET

  MILL SQUARE--FINISH PASS--RPM 880
----------------------------------------------------------------------------------------------------
 0 63 G80 X&  70000 Y-  70000 Z      0              F 140 R  43500 S 12 B 90000 W 1 M17$  253   880    .082
 N 64 G79 X&  63750 Y-  63077                       F 200                             $  259   880    .046
 N 65         Y&  63750                                                               $  261   880    .634
 N 66     X-  63750                                                                   $  263   880    .637
 N 67         Y- 63750                                                                $  265   880    .637
 N 68     X&  63750                                                                   $  267   880    .637
----------------------------------------------------------------------------------------------------
  MACHINING TIME   27.568 MINUTES              TAPE LENGTH   12.09 FEET
```

```
71.284          C I N A C 1  156000 / A P T    I N C H    P O S T P R O C E S S O R   LEVEL A        PAGE    3
```

```
        CPG COMPOSITE TEST PART FOR THE CIM-X 720 WITH TOOL COMP
  O/N  G    X         Y        Z       I       J        F        R       S     B      W   M $ CLNO   RPM    TIME

  MILL TAPERD SQUARE -- FINISH PASS--RPM 880
 N 69 G80                                           R  40375                          $  283   880    .001
 0 70 G80 X&  70000 Y-  70000 Z      0              F 200 R  40375 S 12 B 90000 W 1 M17$  283   880    .004
 N 71 G79 X&  62563 Y-  62694                                                         $  287   880    .052
 N 72     X&  62438 Y&  67562                                                         $  289   880    .626
 N 73     X-  62562 Y&  62438                                                         $  291   880    .624
 N 74     X-  62438 Y-  62562                                                         $  293   880    .624
 N 75     X&  62562 Y-  62438                                                         $  295   880    .624

  MILL CIRCLE--FINISH PASS--1 PASS--RPM 880
 N 76 G80                           Z      6250                                       $  305   880    .005
 0 77 G80 X&  31358 Y-  70000 Z      6250           F 300 R  30000 S 12 B 90000 W 1 M17$  305   880    .016
 N 78 G79                           Z-     6250                                       $  305   880    .020
 0 79 G79 X&  31358 Y-  70000 Z      6250           F 160 R  30000 S 12 B 90000 W 1 M17$  315   880    .039
 N 80     X&  24532 Y-  54761                                                         $  318   880    .143
 N 81 G03 X&  60005 Y&      0        I&      0 J&      0                               $  324   880    .431
 N 82 G03 X-      0 Y&  60005        I&      0 J&      0                               $  324   880    .589
 N 83 G03 X-  60005 Y-      0        I&      0 J&      0                               $  324   880    .589
 N 84 G03 X&      0 Y- 60005         I-      0 J&      0                               $  324   880    .589
 N 85 G03 X&  24532 Y-  54761        I&      0 J&      0                               $  324   880    .157

  MILL SQUARE INSCRIBED IN CIRCLE--FINISH PASS--RPM 880
 N 86 G80                                                                             $  334   880    .000
 0 87 G80 X&  55438 Y-  38818 Z      0              F 160 R  34125 S 12 B 90000 W 1 M17$  340   880    .019
 N 88 G79 X&  52544 Y-  36801                                                         $  344   880    .022
 N 89     X&  36790 Y&  52542                                                         $  346   880    .567
 N 90     X-  52542 Y&  36790                                                         $  348   880    .566
 N 91     X-  36790 Y-  52542                                                         $  350   880    .566
 N 92     X&  52542 Y-  36790                                                         $  352   880    .566

  LOAD TOOL 2 -- 59/64 DRILL
 0 93 G80 X&  52542 Y-  36790 Z      0              F 160 R      0            W 1 M06$  367   880    .117
 0 94 G81 X&  43750 Y-  43750 Z- 11250              F 180 R  36250 S 14 B 90000 W 2 M13$  367  1400    .091
 N 95 G80                                           R      0                          $  369  1400    .018
 N 96 G81 X-  43750                                 R  36250                          $  371  1400    .129
 N 97 G80                                           R      0                          $  373  1400    .018
 N 98 G81         Y& 43750                          R  36250                          $  375  1400    .129
 N 99 G80                                           R      0                          $  377  1400    .018
 N100 G81 X&  43750                                 R  36250                          $  379  1400    .129

  LOAD TOOL 3 -- 3/4-14 PIPE TAP
 0101 G80 X&  43750 Y&  43750 Z      0              F 180 R      0            W 2 M06$  393  1400    .118
 0102 G84 X-  43750 Y-  43750 Z- 10000              F  30 R  36250 S  3 B 90000 W 3 M13$  393   175    .746
 N103 G80                                           R      0                          $  395   175    .018
 N104 G84 X&  43750 Y&  43750                       R  36250                          $  397   175    .746

  LOAD TOOL 4 -- 1.0 BORE(CARB. TIP)
 0105 G80 X&  43750 Y&  43750 Z      0              F  30 R      0            W 3 M06$  409   175    .118
 0106 G85 X&  43750 Y-  43750 Z- 5000               F  60 R  36250 S 15 B 90000 W 4 M13$  409  1755    .228
 N107 G80                                           R      0                          $  411  1755    .018
----------------------------------------------------------------------------------------------------
  MACHINING TIME   37.665 MINUTES              TAPE LENGTH   20.10 FEET
```

FIGURE 9-7

Continued...

```
 71.284            C I N A C 1  156000 / A P T    I N C H    P O S T P R O C E S S O R   LEVEL A        PAGE    4
------------------------------------------------------------------------------------------------------------------
     CPG COMPOSITE TEST PART FOR THE CIM-X 720 WITH TOOL COMP
     O/N   G       X          Y          Z        I        J        F      R      S      B     W   M $ CLNO   RPM    TIME
     N108 G85 X-  43750 Y&  43750                                        R  36250                      $  413 1755    .246
------------------------------------------------------------------------------------------------------------------
     LOAD TOOL 5 -- 5/16 DRILL
     O109 G80 X-  43750 Y&  43750 Z       0                              F  60 R       0               W 4 M06$  425  1755   .118
     0110 G81 X&  25000 Y-  25000 Z    8750                              F 120 R  30000 S 17 B 90000 W 5 M13$  425  2785   .140
     N111     X-  25000 Y&  25000                                                                       $  427  2785   .112
------------------------------------------------------------------------------------------------------------------
     LOAD TOOL 6 -- 3/8-16 N.C. TAP
     0112 G80 X-  25000 Y&  25000 Z       0                              F 120 R       0               W 5 M06$  439  2785   .114
     0113 G84 X&  25000 Y-  25000 Z   3750                              F  30 R  30000 S  4 B 90000 W 6 M13$  439   220   .285
     N114     X-  25000 Y&  25000                                                                       $  441   220   .285
     N115 G80 X&  150000 Y&  60000                                                                       $  445   220   .089
     N116                                                                                               M02$  447   220   .000
     LEADER/   72.0
------------------------------------------------------------------------------------------------------------------
     CPG COMPOSITE TEST PART FOR THE CIM-X 720 WITH TOOL COMP
------------------------------------------------------------------------------------------------------------------
     MACHINING TIME    39.074 MINUTES                       TAPE LENGTH   22.32 FEET


                          ** END OF POST PROCESSING **

                       ELAPSED TIME IS    0.36300 MINUTES
```

FIGURE 9-7
Continued...

CAD/CAM YESTERDAY AND TODAY

As computer-aided design/computer-aided manufacturing (CAD/CAM) systems began to make a profound impact on the engineering/manufacturing process in the mid to late '70s, hardware elements underwent considerable change and cost reductions, and software increased in capability and functionality. Cost reductions in the CAD/CAM industry were brought about by the same factors that decreased costs in the CNC industry: smaller and more powerful integrated circuits and microprocessors and increased computer power. In the 1980s, costs further decreased as computing and graphics functions became tightly coupled in mainframe applications, and engineering workstations and personal computers (PCs) became the dominant computing platform. The result was mainframe power on a PC or workstation for a fraction of the mainframe cost. Graphics terminals have now become smaller and more powerful, color displays have become standard, and very powerful programming techniques for manipulating computerized images have been developed.

Earlier systems (see figure 9-8) were monochromatic and basically large, expensive, and slow; they used large storage-tube and vector-stroke technology. Additionally, they required proprietary operating systems and hardware. However, CAD systems were electronic substitutes for conventional drawing boards and represented a vast improvement over conventional design and drafting techniques. Additionally, up to 40 or 50 terminals could be connected to mini and mainframe computers. This computer sharing was intended to maximize the use of a critical computing resource but often resulted in inconsistent and slow response time. Current systems can run on a variety of hardware platforms (although some are still single-platform based); they are color, compact, relatively inexpensive and high-performance, and can store all the CAD data electronically.

FIGURE 9-8
Early CAD/CAM systems
were monochromatic,
basically large, expensive,
slow, and mainframe
based. (Courtesy of
Cincinnati Milacron Inc.)

CAD/CAM systems allow for component parts and assemblies to be designed in an interactive environment, with design geometry stored in a central database for access and retrieval. The simplest CAD systems are two-dimensional (2D) and contain functional architecture for the design and drafting aspects of component parts and assemblies on a computer graphics CRT (cathode ray tube) display terminal. The vast majority of CAD use is 2D. CAD systems are primarily used to create lines, surfaces, solids, intersections, and curved surfaces. In simpler terms, they create, transform, and display pictorial, descriptive, and symbolic data.

CAD capabilities range from using computers to create drawings to performing isolated calculations and compiling a bill of materials. Recorded images can range from a simple straight line to a multicolored pictorial representation of a three-dimensional (3D) assembly. Some images feature sculptured surfaces and moving parts, with shading and perspective to promote depth visualization. These descriptive geometric representations can then be rotated and viewed like an object in space, giving the designer total part-viewing capabilities.

CAD systems are highly interactive and user oriented, as shown in figure 9-9. In many cases, they use construction techniques familiar to the conventionally trained draftsman. An interactive system provides immediate responses to the user's instructions, changes, or additions through menu selection. This greatly enhances a designer's productivity, creativity, and conceptual thinking. Before CAD, the designer would sit in front of a drawing board; now he or she sits in front of the graphics terminal or workstation and creates geometric images on the screen. The designer can also add and reproduce dimensions and symbols and manipulate the constructed images in a variety of ways never before possible with conventional paper and pencil. Once the engineer/designer has arrived at the final version or design, the image—because it is based on mathematical coordinates and entities stored within the computer—can be transmitted to peripheral devices for printing, plotting, etc.

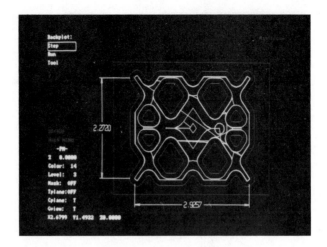

FIGURE 9-9
A modern CAD system in use.
(Courtesy of CNC Software Inc.)

One of the most important aspects of a CAD/CAM graphics system is that once the final part image is created, it can be stored in the engineering computer database. This makes it readily accessible for viewing by other engineering personnel and ultimately for manufacturing use. From a design point of view, the storage of images greatly enhances and aids compatibility and interference visualization. Views can be merged, stacked, and rotated for assembly clarification without anyone having to draw the assembly on paper and then search for interference factors that necessitate redesign. Replication of details is also possible. A designer may construct such details as a fastener or a bracket only once and then replicate and locate it as necessary, making the geometry of a part available to other users. A library of standard symbols can be stored in the system and called up by users as needed. Most engineering database systems provide both data management and data protection. The system controls the deletion of data and protects against unauthorized changes to drawings. Terminal and use activity can be monitored and recorded on a regular basis.

In addition to providing interactive design geometry capabilities, most CAD systems also provide advanced and powerful software programs that can analyze and test a design before any prototype parts are manufactured. Internal routines, such as finite element analysis, allow the engineer to calculate and predict patterns of stress and strength as well as other critical factors such as volume and weight (see figure 9-10*).

As CAD/CAM systems continued to evolve, it became obvious that no standards existed for how geometric information was structured and stored. Consequently, geometric data could not be exchanged from one system to another. Early efforts involved creating one-to-one translators between systems, but that proved unmanageable because multiple translators would be needed for every one-to-one combination of systems. Eventually, a standard data format with a neutral database for geometric translation was proposed in 1979 by the U.S.

---

*See also color insert in Chapter 11.

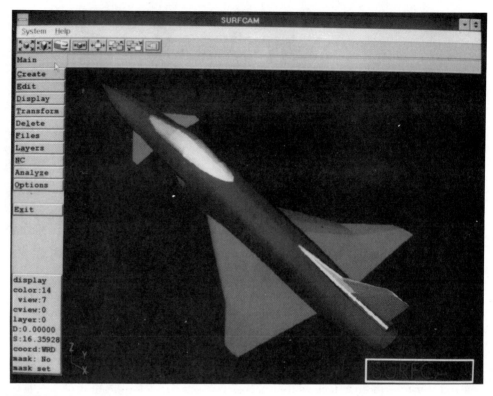

FIGURE 9-10
CAD systems also provide powerful programs such as finite element analysis to test and predict patterns of stress and strength in addition to calculating volume and weight. (Courtesy of Surfware Inc.)

Air Force. This standard data format, known as the Initial Graphical Exchange Specification, or IGES, was supported by other Defense Department groups and CAD/CAM users and suppliers; it became the standard in 1981. Most suppliers now provide an IGES-in and IGES-out capability for their customers.

With today's advanced systems, 3D surface and solid models can be created. A surface model is one in which the surfaces, edges, and primary curves of the model are defined. The appearance is that of a solid object, as in figure 9-11. Surface models are usually used for N/C. A solid model freely describes the edges and faces of the design and also provides the knowledge of how the edges and faces are connected (topology). Models done in 3D can be used to produce 2D drawings.

Graphics creation of a design image is done in three phases. The design or model is first converted from an electronic format to a precise visual form. Second, the display-format image is stored and then mapped to a display. Finally, the image is displayed on the screen. Solid modeling provides the best and most complete representation of an object, but extensive workstation or PC computing power is required to handle the workload requirements.

FIGURE 9-11
With today's CAD/CAM systems, accurate surface
and solid models can be created, such as this
example of an impeller (see part A), leading to
faster and better completion of the finished part
(see part B). (Courtesy of CGTech Inc.)

(Part B: Finished Part)

In addition to the basic CAD and CAM functions common to all CAD/CAM
systems to one degree or another, most major mainframe CAD/CAM vendors
have systems with the following additional features:

- Modeling for manufacturing geometry creation
- Model-merging techniques
- Label and dimension creation
- Customizable and configurable user-interface
- On-line help documentation
- Fixture, part, turret, and tool assembly modeling for tool path verification
  and interference checking
- Techniques for controlling sequence of N/C output
- Curve-fitting capability
- Tool and machining libraries
- Full two-axis lathe turning, grooving, drilling, and threading applications
  for tool path creation
- APT macro integration into tool path
- Tool path verification and modification, including multiaxis rough and
  finish milling
- 2.5-axis hole drilling, with optimization and cycle display capability

- Associative tool path transformations
- Customizable, pop-up, on-screen menus
- Full interactive control for adding APT or postprocessor statements

One CAD/CAM feature that deserves special consideration is *associativity*. Full associativity between part geometry and interactive machining allows each tool path to be automatically changed to reflect changes made to a single geometry model. Essentially, when the design model is changed, the N/C tool path is automatically altered to suit the changed geometry. The cutter path, however, must still be regenerated and repostprocessed. This is a tremendous productivity and time-saving improvement over conventional language-based programming, for which definition and possibly tool motion statements would have to be changed and the program rerun and repostprocessed every time a design engineer changed the part drawing. Additional time is saved because the part geometry changes are now electronic; previously, such changes would have required an engineering change notice to be issued and new part drawings to be done for every change.

GRAPHICS-BASED N/C PROGRAMMING

Although language-based processors provided a big productivity boost over manual programming methods, their continued use and development presented several distinct problem areas:

1. Detailed part geometry, created electronically via CAD and stored in an engineering computer database, had to be recreated using APT or other language definition statements. This is redundant. Why recreate part geometry that already exists?

2. Language-based processors are text-only—no graphics. Producing an accurate program depends on the programmer's ability to conceptually "visualize" the machining process while writing tool motion statements. Verifying and checking for tool path accuracy, fixture interferences, and other machining related problems could only be accomplished through close mathematical checking of the postprocessor and CL output, plotting, and tape prove-out, or dry run, on the machine.

3. The base of skilled, trained programmers knowledgeable in APT and other processor languages was diminishing.

4. Language-based programming is entirely too time consuming and error-prone. Manufacturers must reduce the overall product development, manufacturing, and assembly cycle time to bring products to market more quickly.

Interactive graphics-based systems were introduced to greatly simplify the programming process, move visualization from the programmer's mind to the computer screen, and provide the opportunity for immediate feedback to the programmer. These interactive systems, as shown in figure 9-12*, make

---

*See also color insert in Chapter 11.

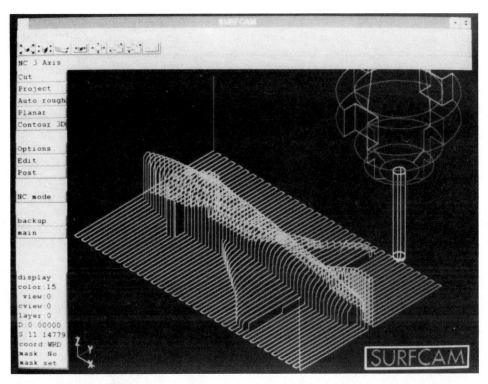

FIGURE 9-12
Interactive graphics-based system. (Courtesy of Surfware Inc.)

use of color graphics, multitasking, pop-up menu displays that can design a part in three dimensions, create detailed part prints, design tools and fixtures, and write the N/C program to produce the part—all with the same system. Graphics-based systems, many of which are PC based, use a mouse as a pointing device, along with easy-to-understand menu-guides, to enter and change geometry and tool motion. Users can easily define part geometry, create the tool path, reduce errors by visualizing results, and greatly improve productivity through the interactive capability to reprogram and replot. Also, the majority of N/C verification can be accomplished on the graphics display terminal, rather than idling the N/C machine and operator on the shop floor for tape prove-out.

To describe a part, most N/C programming systems provide their own modeling capability. The CAD-like "front end" permits users to create a part by drawing lines, circles, arcs, and splines. Interactive graphics help N/C programmers create the tool path by defining the exact path the tool will take to cut the part. N/C programming using graphics enhances the part programmer's ability to visually follow the tool path by showing a three-dimensional view of cutter clearance planes, retract planes, depth planes and clamps, fixtures, and casting clearances. An interactive graphical tool-path-editing capability, now available in some N/C packages, provides the N/C programmer with an additional and practical technique to modify tool paths that have been generated. Calculated tool paths are then verified, and the computed coordinates are transmitted to a post-processor for N/C output data generation. Thus, an N/C programmer can sit in front of a graphics display terminal (just like the designer), create the tool paths, and then compare the calculated and verified tool paths with the actual part.

Graphics-based systems must also handle design inadequacies such as surface discontinuities, and many consider the complete cutter and holder geometry in relation to workpiece and check-surface geometry (see figure 9-13*). This feature allows gouge-free tool paths to be generated. As part of the machining strategy, other features allow menu selection of the machine tool, material type, specific tools for each cut, and speeds and feeds.

As the cost of these systems reduces further and they become more widely used, their overall impact on conventional programming methods will become more significant. Such systems will continue to evolve and be slanted toward increased integration, connectivity, and compatibility with other graphics systems in order to share information. Special emphasis will be placed on produc-

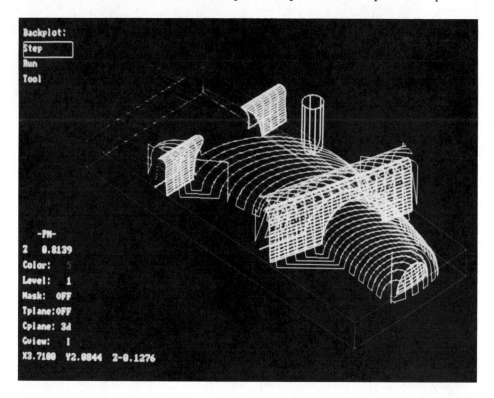

FIGURE 9-13

Some graphics-based systems consider the complete cutter and holder geometry in relationship to the workpiece to generate gouge-free tool paths. (A) Courtesy of CNC Software Inc.; (B) Courtesy of CGTech Inc.

*See also color insert in Chapter 11.

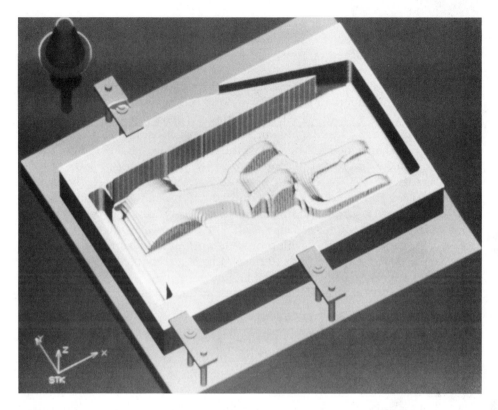

FIGURE 9-13
Continued

ing "redundancy-free" software to minimize user input and take advantage of previously entered and captured data and geometry.

NEW CNC FEATURES

Just like other computers, CNCs contain ROM and RAM memory. ROM, or read-only memory, is written into the electronic chip and cannot be erased or altered without special equipment. ROM can be accessed by the computer but cannot be altered by day-to-day users. This is why the CNCs resident executive program cannot be erased and is always active when the MCU is on. RAM, or random-access memory, can be accessed and altered (written to) by the computer. The CNC code is entered into RAM via either the keyboard or an outside source such as floppy disk or by the downloading of a DNC file to the CNC unit. The contents of RAM are lost when the machine is turned off, so RAM is referred to as *volatile* memory. In contrast, ROM, which is *nonvolatile*, will not be lost when the MCU is turned off.

Many CNC controllers utilize a battery backup system that powers the computer long enough for the program to be transferred (saved) to some storage media in the event of power loss; other CNC controllers use a special type of RAM called CMOS memory, which retains its contents even when the power to the computer is turned off.

The control systems of today do essentially the same things that they did 15 years ago: move slides around and control the spindle and other peripherals. But major improvements in both software and hardware have occurred in the last few years. Many of the new CNC operational characteristics and programming advancements, as well as the incresased user friendliness, are a result of the software enhancements. The cost of the new software development can be as much as 70% or more of the control.

CNC hardware (see figure 9-14*) has also advanced considerably, primarily centered around the development and adoption of the 32-bit microprocessor. This microprocessor has dramatically improved processing speed, which in many cases is at least 10 times faster than its older predecessor, the 16-bit control. A very good 16-bit CNC will process a block of new coordinate values in three axes in 50 to 60 milliseconds (thousandths of a second). A 32-bit processor can process a block in as little as 2 milliseconds.

Other features of typical 32-bit processors include the following:

- 75,000 ft. of part program storage
- 20-MB hard disk
- Improved graphics and diagnostics
- Tool and part management
- Enhanced communications capabilities
- Multitasking operating system, which permits the machine tool to operate at maximum capability while the operator performs other CNC tasks such as program downloading and tool data updates
- Improved graphics capability that allows the programmer to see the outcome of his or her work on the CRT screen before any metal is cut or tools broken

Microprocessors, once used sparingly in CNCs, are growing in terms of new capabilities and tasks to be performed at the shop floor level, while decreasing in size and cost. Additionally, they are simpler and easier to use, with much higher reliability. In some cases, CNC manufacturers have developed custom-designed integrated circuits known as application-specific integrated circuits (ASICs) to perform high-speed operations such as axis-interpolation and compensation. Collectively, the parts of a CNC that work together to provide improved machine tool performance include the following:

1. CPU
2. Electro/mechanical servo control systems for spindle and axis control
3. Various machine, human, and auxiliary interfaces
4. The internal bus structure, or information highway, which passes information from one portion of the control system to another

---

*See also color insert in Chapter 11.

FIGURE 9-14
The development and adoption of the 32-bit microprocessor has dramatically
increased the processing speed and computational power of the CNC, by
10 times or more over the older 16-bit processor. (Courtesy of Cincinnati
Milacron Inc.)

SHOP FLOOR PROGRAMMING

Shop floor programming (SFP) is an automated part programming system
embedded in a CNC, with powerful capabilities equal to many commercially
available, off-line, personal computer–based programming systems. Unlike the
hard-wired N/C versions of the 1950s and 1960s, CNC in the 1990s does not require significant training. Programming input can be in conversational English.

The shop floor programming feature of a CNC greatly simplifies the task
of programming directly from a part drawing. Even a relatively inexperienced
programmer can obtain immediate results. SFP provides the user with the
capability to run in background mode and create a new program or modify an

existing program right at the machine while the machine is busy cutting a previously programmed part. The operator/programmer can enter the material name and part-shape dimensions, and the system will automatically determine tool types, speeds, feeds, and cutting tool order. In the programming mode, the CRT prompts the programmer through the entire programming process and displays the results. Then it produces a complete program. Cutter path instructions are generated, along with all of the pertinent preparatory and miscellaneous codes, functions, and commands—all transparent to the user. After any necessary edits are performed, the programming system simulates and displays the machining process on the CRT screen (see figure 9-15) while checking for any possible interference.

Normally treated by CNC manufacturers as an optional feature, SFP is offered by some as a standard part of their CNC package. On some SFP systems, the display may be only the path plot of the cutter's centerline, whereas on other systems there may be a very elaborate graphics display showing an image of the part, the cutter path, the cutter, and even the simulation of stock removal. SFP should not be confused with manual data input (MDI), which is strictly a manual means of inputting block-formatted data and is a standard feature on all CNCs.

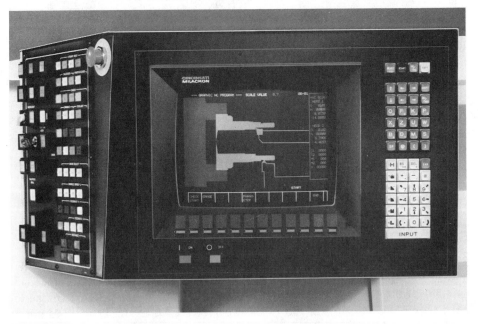

FIGURE 9-15
The SFP feature of a CNC provides the capability not only to create a new program or modify an existing program at the machine, but also to simulate and display the machining process on the CRT screen. (Courtesy of Cincinnati Milacron Inc.)

REVIEW QUESTIONS

1. Name the four main sections of APT processing, and briefly explain their function.

2. What are the major functions of a postprocessor? What is a postprocessor generator?

3. What is BCL, and what are its advantages over conventional N/C postprocessing?

4. What are the three major elements of APT part programming?

5. Explain why processor languages such as APT continue to be used.

6. Describe some of the newest CAD/CAM developments and enhancements over their earlier predecessors.

7. List some CAD/CAM features.

8. Briefly discuss what is meant by associativity.

9. Define IGES, and discuss why it was developed.

10. What single major development led to the improved processing speed, increased computing power, and advanced functionality of a modern CNC unit?

11. Explain the difference between ROM and RAM memory.

12. List some new CNC features.

13. What major components of a CNC work together to provide improved machine tool performance?

14. Explain the application and use of shop floor programing (SFP).

CHAPTER 10

Tooling for N/C and CNC Machines

OBJECTIVES _____

After studying this chapter, you will be able to:

- Describe the overall importance and impact of proper tooling on N/C and CNC machines
- Explain how correct use of cutting tools affects overall machine performance and productivity levels
- Discuss how proper fixturing leads to successful use of an N/C or CNC machine
- Identify some sound tooling practices for productive part-processing

TOOLING CONSIDERATIONS

Tooling for N/C and CNC machines has always been one of the most neglected elements of an N/C installation. During the planning and justification stages for the purchase of N/C equipment, this aspect is often given secondary consideration because all tooling tends to be taken for granted until something goes wrong.

N/C machines can only move or position appropriate cutting tools to specific locations and rotate them or the workpiece at desired spindle speeds. The individual cutting tools actually do the metal removal work. The only way an N/C machine can be efficiently and effectively used is through proper use and care of cutting tools and work-holding devices.

In conventional machining, part accuracies depend on special fixturing. This type of fixturing has precisely made and located tool-setting pads and accurately located bushings that guide the tools. With an N/C machining center, simple fixturing is used. There are no tool bushings or tool pads to guide the tools. The repetitive positioning accuracy of the machine promises a high degree of quality. However, machining accuracies depend on the inherent accuracies of the cutting tools and their holders. If a drill "runs out," the benefit of the machining center's accuracy is lost. The programmer must assume that the tools will not run out.

There is another reason for careful selection of cutting tools. The average, conventional machine tool cuts metal only 20% of the time. An N/C machining center can be expected to cut metal up to 75% or more of the time. This results in more tool use in a given period of time. Tool life, measured in terms of "time in the cut," will be as good or better, but because of the increased use, cutting tools will be used up three times as fast. The cost of perishable tools used during the machine's lifetime may amount to 50% or more of the machine's purchase price. Therefore, perishable tools represent a sizable investment—hence the importance of getting high volume along with good tool life.

It should be noted that an N/C machine is no more accurate than the cutting tools used with it to machine the workpiece. Thus, the decision of which cutting tools and toolholders to buy should receive the same consideration as was given the purchase of the machine.

CUTTING TOOLS USED ON N/C EQUIPMENT

A variety of cutting tools are used on N/C equipment to perform a multitude of machining operations. Many of the cutting tool applications, however, are no different than those that would have to be performed on manual equipment to produce the same workpiece. Cutting tools range from conventional drills, taps, and end mills to high-technology carbide cutting tools. Because of the importance of cutting tools to the overall manufacturing process, as well as the tools' costs, it is important that each be examined in detail.

Drills

Even though the slide-positioning accuracy of most modern machining centers is ± 0.001 in. or better, there is no guarantee that the drilled hole location will be within that degree of accuracy. A standard, commercial twist drill, manufactured to specifications, may be very accurate; or it may be so inaccurate that nothing more than roughing work is possible.

All new drills have certain allowable tolerances, as depicted in figure 10-1. Those tolerances that affect accuracy the most are lip height, web centrality, and flute spacing. The lip height, for example, of a 0.250-in. drill can vary

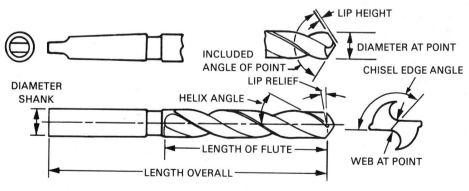

FIGURE 10-1
Identification of toleranced areas for two-flute general purpose drills.

0.004 in., its web can be off center as much as 0.005 in., its flute spacing can be off by 0.006 in., and the drill will still be within required specifications. Since a 0.250-in. drill is normally fed at a rate of 0.004 or 0.005 IPR, it would be impossible for that drill to produce accurate holes.

Since approximately 70% of all hole making is drilling, tool selection is of primary importance. One of the most important steps in the selection of a drill is to choose the shortest drill length that will permit drilling of the hole to the desired depth. A good rule to remember is: the smaller the drill size, the smaller the allowable error; as drill size increases, the allowable error progressively increases. Short, stubby drills run truer, allow the fastest feeds, and improve tool life. The torsional rigidity of a drill will affect not only tool life and feed potential, but hole quality as well. Torsional rigidity is a measure of the tool's ability to resist twisting or unwinding; rigidity increases as drill length decreases. Therefore, on machining centers, where feed is constant and rapid, a shortened flute becomes a distinct advantage.

Many different types and varieties of drills exist and are used for a wide variety of applications. The common twist drill certainly has its applications, but so do center drills, spade drills, and subland drills. Center drills, or spotting drills, are primarily designed to produce accurate centers in the work so that follow-up drills will start in perfect alignment. The proper selection and use of these drills will increase the accuracy of hole location, particularly on rough surfaces.

Ideally, the center drilled hole should be machined to a depth where the countersunk portion is 0.003 to 0.006 in. larger than the finished hole size (see figure 10-2). With this method, the drill periphery will be guided into the countersunk hole, the location will be accurate, and the finished hole will have a chamfered or deburred edge.

The most widely used center drill is the bell, or combination, type (see figure 10-3). It is commonly used in lathe work to provide work centers for subsequent

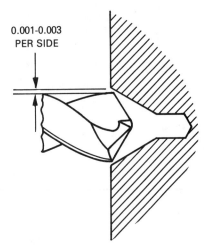

0.001-0.003 PER SIDE

FIGURE 10-2
Ideal center drill size in relation to finish hole size.

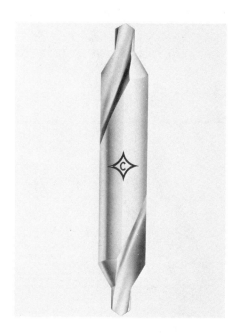

FIGURE 10-3
Bell, or combination, center drill. (Courtesy
6f Cleveland Twist Drill)

operations. The advantages of this drill are accuracy and availability. However, for work assigned to an N/C machine, it has two disadvantages. First, the lead portion of the drill breaks off quite easily. Second, the drill is limited to small diameters.

If large-diameter holes are required to be machined to relatively accurate tolerances, the twist drill may be impractical. Spade drills are sometimes considered (see figure 10-4) because they can produce large holes in one pass. In contrast, there is the conventional twist drill, which makes progressively larger holes until the desired size is obtained.

Spade drills are advantageous in N/C work because only the blade, not the entire tool, needs to be changed when it becomes dull. Thus, correct tool length is maintained, and resetting the tool length or recompensating the machine is eliminated.

The spade drill will normally use the same feeds and speeds as a twist drill. In cast iron, the spade drill performs well at almost any depth. However, in steel and aluminum, if the hole depth is more than 1.5 or 2 times the hole diameter, problems with heat and chip removal can occur.

Price is another important consideration in drill selection and purchase. Generally, on a range of drill sizes from 1 to 2 in. (by $\frac{1}{32}$ of an inch), twist drills cost twice as much as spade drills. This is mainly because various blade sizes are interchangeable in a single shank. For example, only three spade drill shanks are required to hold the entire 1 to 2-in. blade range.

Many hole-producing sequences require multiple operations on the same hole, such as drill and countersink, drill and counterbore, or drill and body drill. Multiple-diameter, multiple-land tools, called subland tools, are commonly used

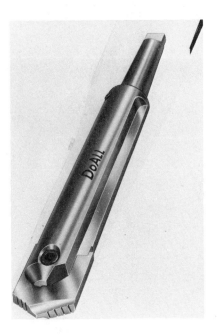

FIGURE 10-4
Spade drill, showing blade and shank.
(Courtesy of DoAll Company)

today (see figure 10-5). The proper use of this type of drill can result in a time savings and quality improvement. By combining multiple-drilling operations into one tool, extra machining time and tool-handling time are eliminated. An additional benefit is derived from the rigidity of the larger diameter, both in the ability to use maximum feed rates and in improved hole accuracy.

FIGURE 10-5
A multiple-diameter (subland) drill.
(Courtesy of Cleveland Twist Drill)

Taps

Tapping is one of the most difficult machining operations because of the ever-present problem of chip clearance and adequate lubrication at the cutting edge of the tap. This is further aggravated by coarse threads in small diameters, long-thread engagements, unnecessarily high thread percentages, tough materials, and countless other factors. Further, the relationship between speed and feed is fixed by the lead of the tap and cannot be independently varied. Tapping on an N/C machine (rather than manually) requires less operator skill to tap a good hole because the tapping operation is programmed. Therefore, the prime concern is the tap, not the skill level of the operator.

Generally, taps are divided into two major classifications: hand taps and machine screw taps. Their names, however, do not denote the manner in which the taps are used, because they are both used in power tapping of drilled holes.

Hand taps (see figure 10-6) were originally intended for hand operation, but they are now widely used in machine production work. The name denotes the group of taps that are available in fractional sizes. The most commonly used hand taps include sizes ranging from $\frac{1}{4}$ to $1\frac{1}{2}$ in.

Machine screw taps refer to the group of taps available in decimal sizes. Machine screw taps are actually small hand taps. Their size is indicated by the machine screw system of sizes, ranging from #0 to #20.

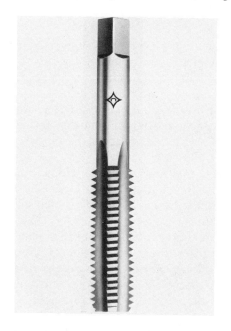
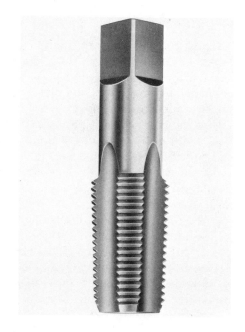

A   B

FIGURE 10-6

Conventional hand taps. (A, Courtesy of Cleveland Twist Drill; B, Courtesy of Morse Cutting Tools Division, Gulf & Western Manufacturing Company)

FIGURE 10-7
Spiral pointed (or gun or chip-driver) tap.
(Courtesy of DoAll Company)

Spiral pointed taps (see figure 10-7) are sometimes referred to as gun, chip-driver, or cam point taps. They are recommended for through holes and for holes with sufficient clearance at the bottom to provide chip space. Spiral pointed taps have straight flutes with a secondary grind in the flutes along the chamfer. This secondary grind is ground into the flute at an angle to the axis of the tap so that it produces a shearing action when cutting the thread. As a result of this shearing action, the chips are forced ahead of the tap with very little resistance to thrust.

The main advantage of the spiral pointed tap is that it prevents chips from packing in the flutes or wedging between the flanks and the work. This is a major cause of tap breakage, particularly in small taps. The spiral pointed tap also allows a better flow of lubricant to the cutting edges.

Spiral-fluted taps (see figure 10-8) are recommended for the tapping of blind holes, where the problem of chip elimination is critical. They are most effective when the material being tapped produces long, stringy, curling chips. The spiral-fluted tap cuts freely while ejecting chips from the tapped hole. This prevents clogging and damage to both the threads and the tap. Chip removal is accomplished by the backward thrust action of the spiral flutes.

In tapping operations, there are many factors that reduce tap life. Studies of the relevant factors indicate that most problems stem from one area: eliminating chips from the hole. However, with fluteless taps (see figure 10-9), this problem and others can be solved. Fluteless taps do not produce chips. Rather, they form or roll the threads into the hole through cammed lobes

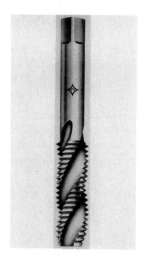

A

B

FIGURE 10-8
Spiral-fluted, or turbo, taps. (A, Courtesy of Cleveland Twist Drill; B, Courtesy of DoAll Company)

FIGURE 10-9
Fluteless taps. (Courtesy of DoAll Company)

on the periphery of the tap. Because of this forming action, the tap drill used is always larger than that used for a conventional tap. Fluteless tapping requires one deviation from normal tapping practice. Since it is basically a forming rather than cutting operation, a high-pressure lubricant should be used rather than a cutting compound. This is extremely important when fluteless taps are used to tap the tougher materials.

Traditionally, certain materials have been tapped dry—plastics and cast iron, for example. Even where dry tapping is possible, improved performance is generally effected by a judicious selection of some type of lubricant. Logically, lubrication is one of the key elements of successful tapping. It must therefore receive more than casual attention.

Tapping lubricants serve several purposes, the most important being the following:

- They reduce friction.
- They produce clean, accurate threads by washing chips out of the tap flutes and the threaded hole.
- They improve the thread's surface finish.
- They reduce build-up on edges or chip welding on the cutting portion of the tap.

In summary, when purchasing a tap, you are in effect buying tapped holes rather than taps. Thus, tap life becomes the primary concern.

Reaming

Reaming is the process of removing a small amount of material, usually 0.062 in. or less, from a previously produced hole. The reamer is a multibladed cutting tool designed to enlarge and finish a hole to an exact size. Since the reamer is basically an end cutting tool mounted on a flexible shank, it cannot correct errors in hole location, hole crookedness, etc. A reamer will follow a previously produced hole. Therefore, when straightness or location is critical, some prior operation, such as boring, must have been performed in order to obtain these qualities.

Reamers can have either straight or spiral flutes and either a right- or left-hand helix (see figure 10-10). Those with spiral or helical flutes will ordinarily provide smoother shear cutting and a better finish.

The shell reamer (see figure 10-11) is used primarily for sizing and finishing operations on large holes, usually $\frac{3}{4}$ in. and larger. The reamer will fit either a straight or taper shank arbor. Several different sizes of shell reamers may be fitted to the same arbor. This results in a tool savings.

Reaming on an N/C machine is an accepted practice. Floating toolholders are not usually required for reaming operations on an N/C machining center for two reasons. First, the repetitive positioning accuracy of the machine slides ensures that the reamer will be positioned in line with the hole. Second, the accuracy of present-day collets and adaptors ensures that the reamer will run "true," and will therefore cut to size.

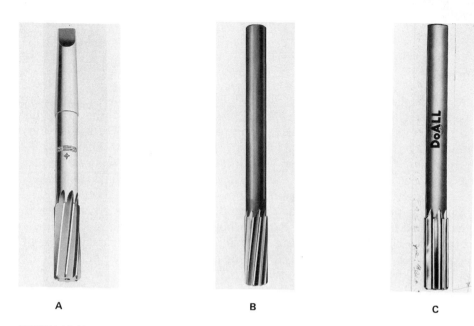

A

B

C

FIGURE 10-10

Some typical reamers: spiral and straight flutes. (A, Courtesy of Cleveland Twist Drill;
B, Courtesy of Morse Cutting Tools Division, Gulf & Western Manufacturing Company;
C, Courtesy of DoAll Company)

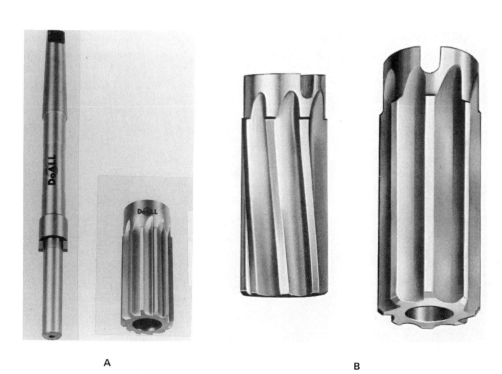

A

B

FIGURE 10-11

Shell reamers. (A, Courtesy of DoAll Company; B, Courtesy of Morse Cutting Tools Division, Gulf
& Western Manufacturing Company)

Boring

Boring is a machining operation that implies extreme accuracies. Therefore, a boring machine should be capable of positioning within tenths of an inch and capable of holding roundness within millionths of an inch. If roundness within two or three tenths of an inch and positioning accuracy within ±0.0005 or ±0.001 in. is satisfactory, then some CNC machining centers can be used as boring machines. Even then, these tolerances are very difficult to maintain without a thorough knowledge of boring methods and boring tools.

Boring operations are performed to produce accurate diameters, accurate locations, good finishes, and true, straight holes. Properly performed, boring is the one hole-finishing process whereby the full positioning accuracy of an N/C machine can be used. A cored hole may be cast out of location, or, in drilling, the drill may wander beyond the acceptable tolerance. Boring can correct these errors and finish the hole with a high degree of accuracy.

A variety of boring bars, similar to the ones in figure 10-12, and numerous types of cutters are used for modern boring operations. Regardless of the type used, all have certain common characteristics. Every boring cutter has a side cutting edge and an end cutting edge. These edges are related to the tool shank and are part of the standard nomenclature of single-point cutting tools of the American National Standards Institute (ANSI).

The type of boring to be done determines, to an extent, the boring tools needed. For example, of the average parts processed on a machining center, 70% are drilling parts, 20% are milling parts, and 10% are boring parts. If boring is only an occasional operation, an offset-type boring head may be used. Even though it has a limited stock-removal capability, the wide adjustment range

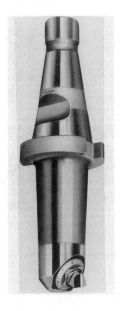

A

B

FIGURE 10-12
Typical boring bars with cartridge inserts. (A, Courtesy of Cincinnati Milacron Inc. B, Courtesy of Kennametal Inc.)

is a definite advantage. If boring requirements are on more of a production basis and cost must be kept to a minimum, then cartridge-type cutters are more economical. Turning centers also use a wide variety of boring bars and inserts. Consult local tooling vendors for additional, up-to-date boring bar and cutter information.

Another element of boring that deserves consideration is the length-to-diameter ratio (L/D). This ratio is one of the most important, but most neglected and least understood, aspects of boring. It refers to the length of the boring bar in relation to its diameter. Some of the largest manufacturers of boring machines and boring bars have researched this problem extensively. Studies indicate that a boring bar with a 1:1 length-to-diameter ratio is 64 times more rigid than a boring bar with a 4:1 ratio; it is 343 times more rigid than one with a 7:1 ratio. It follows, then, that in order to obtain maximum rigidity and accuracy from boring operations, the boring bar should be as short as possible.

Milling

With the exception of drills, probably the most widely used and efficient metal removal tool for a machining center is the end mill. Although arbor-mounted milling cutters on long production runs produce cheaper chips per dollar's worth of tool, the end mill is usually more economical for job shop quantities. Since machining centers are most efficient for short- to medium-sized production runs, it is evident that an end mill similar to that in figure 10-13 takes its place as one of the basic tools for machining centers.

Because of the work potential of end mills, a machining center's contouring capability should be used often. Many parts have milled surfaces, bored holes, recesses, cobores, face grooves, and pockets; all of these operations are relatively easy to perform with the two- or three-axis contouring and circular interpolation available on most modern machining centers. In many cases, bored holes can be rough and semifinish bored by the programming of an appropriate end mill in a circular path around the centerline of the bore. This method can result in some tangible savings. One end mill can replace two or more boring bars, and one end mill can be used for several different bores. Tool drum storage space can be freed for additional tools, tool inventory can be reduced, and some time can be saved.

Many other types of end mills may be used on N/C and CNC machines, such as shell end mills with serrated and indexable blades, and face mills for a variety of applications.

FIGURE 10-13
Double-end end mill. (Courtesy of Sharpaloy Division, Precision Industries, Inc., Centerdale, R.I.)

Countersinking and Counterboring

Countersinking on N/C machines can be a frustrating experience because of the difficulty in establishing set lengths. To set up a job accurately for a standard, single-flute, nonpiloted countersink (see figure 10-14), optical measuring equipment must be used. This is mostly because the physical point of the countersink can be from 0.005 to 0.020 in. short of the countersink's theoretical angle vertex. For example, the programmer will calculate Z-axis travel by setting the length of travel based on the countersink's theoretical vertex. Consequently, if the tool is preset to length with an indicator on the physical point, the vertex will be too deep. This will result in an oversized countersink.

As mentioned earlier, countersinking can be a programming problem on an N/C machine. It is important to remember that, when various countersinks are being calculated and programmed, there is a difference between the theoretical vertex and the actual tool point.

Counterboring operations are typically done with a three- to eight-fluted counterbore (see figure 10-15). These tools generally have about a 10° helix and are available with either tapered shanks or small-diameter shanks ideally suited for straight-collet toolholders. Counterbores are designed with fixed or removable pilots to produce counterbores concentric with the drilled hole. With the repetitive positioning accuracy of modern CNC machines, however, the need for piloted counterbores is often eliminated. Except in long-reach applications, standard counterbores may be removed from the tooling of machining centers.

Most shops with machining centers will probably have a stock of end mills, the diameters of which will range from $\frac{3}{16}$ to 2 in. With no need for pilots, it is good machining practice to make the necessary counterbores from end mills. Doing so will yield a large variety of counterbores with almost infinite size availability, and the faster helix means greater feed rates and faster chip removal.

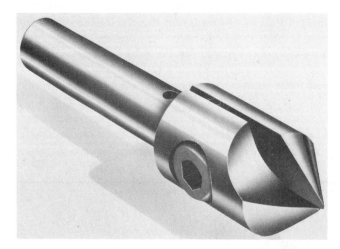

FIGURE 10-14
Standard, single-flute, nonpiloted countersink. (Courtesy of Sharpaloy Division, Precision Industries, Inc., Centerdale, R.I.)

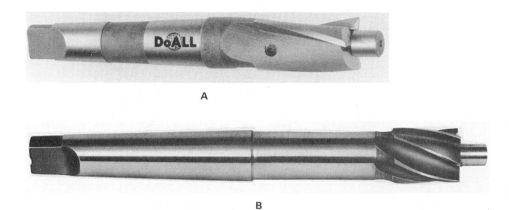

FIGURE 10-15
Typical piloted counterbores. (A, Courtesy of DoAll Company; B, Courtesy of Cincinnati Milacron Inc.)

Some advantages of using end mills as counterbores are reduced tool inventory, lower tool cost per piece, and ease in producing a spotface or counterbore on a rough or angled surface.

CARBIDE INSERT TOOLING

No discussion of N/C tooling is complete without discussing carbide insert tooling. Recent tooling advances have provided modern metalworking with carbide inserts of numerous types and styles, along with titanium-coated and ceramic inserts for both lathe and spindle tooling.

Carbide inserts are available in many standard sizes and shapes, affecting both strength and cost, as shown in figure 10-16. Insert size is measured by the largest circle that will fit entirely in the insert, called the I.C. (inscribed circle). A standard insert has an I.C. of $\frac{1}{4}$, $\frac{3}{8}$, $\frac{1}{2}$, or $\frac{3}{4}$ in. Insert shape is commonly designated by a letter, such as T for triangular, R for round, etc. Inserts also have different nose radii for different applications. Generally, the smaller the I.C. of an insert, the less it costs. The nose radius of an insert affects the surface finish and the maximum allowable feed rate. Large-radius inserts usually leave a better surface finish than small-radius inserts. In addition, large-radius inserts have a stronger cutting edge and dissipate heat better, allowing higher feed rates. Inserts are also identified by the industry-standard identification system shown in figure 10-17. Each position in the insert identification number specifies a position of the insert's geometry.

Carbide inserts are available in a range of grades, each designed for specific machining applications. Hard grades of carbide will contain a higher percentage of tungsten and are more shock resistant. Heavy roughing cuts require a softer shock-resistant grade of carbide, with finishing cuts requiring a harder

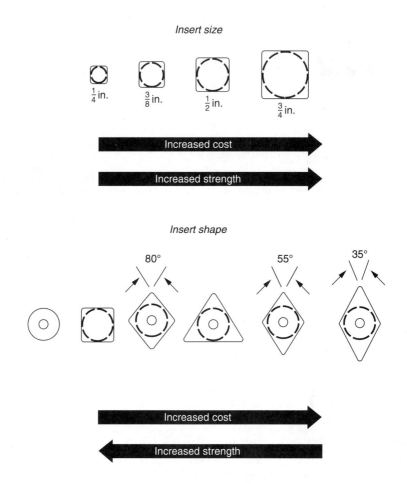

FIGURE 10-16
Insert size, shape, cost, and strength chart.

wear-resistant grade. Additionally, the material being machined influences the choice of insert grade. Nickel-based alloys are very tough and need a soft grade of carbide, whereas abrasive cast irons usually call for a very hard grade of carbide.

Titanium-nitrite (TIN) coated and ceramic inserts are used for both lathe and spindle tooling. The extremely hard and wear-resistant TIN coating is bonded to a soft, tough carbide base, giving the coated insert the best properties of both hard and soft insert grades. TIN coating, usually yellow or gold in color, allows a single insert to be used over a wider range of applications.

Turning Insert Identification

symbol	insert shape	shape	nose angle (degree)
S	□	square	90
T	△	triangular	60
C			80
D			55
E	◇ rhombic (diamond)		75
F			50
M			86
V			35
W	⬡	hexagonal (trigon)	80
H	⬡	hexagonal	120
O	◯	octagonal	135
P	⬠	pentagonal	108
L	▭	rectangular	90
A			85
B	▱	parallelogram-shaped	82
K			55
R	◯	round	—

① shape

tolerances

IC: theoretical diameter of the insert inscribed circle
T : thickness
B : see figure below

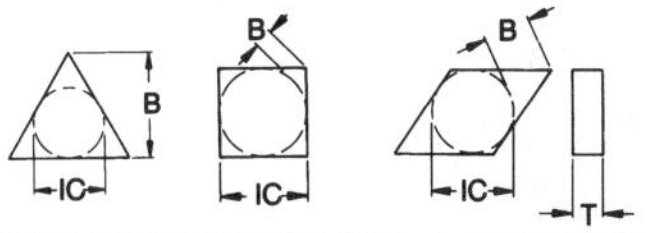

tolerance class	tolerance on "IC"		tolerance on "B"		tolerance on "T"	
	in	mm	in	mm	in	mm
C	±.0010	±0,025	±.0005	±0,013	±.001	±0,025
H	±.0005	±0,013	±.0005	±0,013	±.001	±0,025
E	±.0010	±0,025	±.0010	±0,025	±.001	±0,025
G	±.0010	±0,025	±.0010	±0,025	±.005	±0,13
M	see tables at right		see tables at right		±.005	±0,13
U	see tables at right		see tables at right		±.005	±0,13

③ tolerance

example

inch		C	N	M	G
(metric)		C	N	M	G
position		①	②	③	④

N—0°
A—3°
B—5°
C—7°
P—11°
D—15°
E—20°
F—25°
G—30°

relief angle
②

④ insert type

symbol	hole	shape of hole	chipbreaker	shape of insert's section	alternate symbols ordinary system	IC less than ¼" *
N	without		without		N	
R			single sided		R	E
F			double sided		F	
A			without		A	
M		cylindrical hole	single sided		M	
G			double sided		G	
W		partly cylindrical hole, 40–60° countersink	without		A	
T	with		single sided		M	
Q		partly cylindrical hole, 40–60° double countersink	without		A	D
U			double sided		G	
B		partly cylindrical hole, 70–90° countersink	without		A	
H			single sided		M	
C		partly cylindrical hole, 70–90° double countersink	without		A	
J			double sided		G	
X		special			X	X

*inch system only

⑤ size

inch	IC		metric cutting edge length						
	in	mm	C	D	R	S	T	V	W
1.2 (5)	5/32	3,97	—	04	03	03	06	—	—
1.5 (6)	3/16	4,76	04	05	04	04	08	08	—
1.8 (7)	7/32	5,56	05	06	05	05	09	09	03
—	0.236	6,00	—	—	06	—	—	—	—
2	1/4	6,35	06	07	06	06	11	11	04
2.5	5/16	7,94	08	09	07	07	13	13	05
—	0.315	8,00	—	—	08	—	—	—	—
3	3/8	9,52	09	11	09	09	16	16	06
—	0.394	10,00	—	—	10	—	—	—	—
3.5	7/16	11,11	11	13	11	11	19	19	07
—	0.472	12,00	—	—	12	—	—	—	—
4	1/2	12,70	12	15	12	12	22	22	08
4.5	9/16	14,29	14	17	14	14	24	24	09
5	5/8	15,88	16	19	15	15	27	27	10
—	0.630	16,00	—	—	16	—	—	—	—
5.5	11/16	17,46	17	21	17	17	30	30	11
6	3/4	19,05	19	23	19	19	33	33	13
—	0.787	20,00	—	—	20	—	—	—	—
7	7/8	22,22	22	27	22	22	38	38	15
—	0.984	25,00	—	—	25	—	—	—	—
8	1	25,40	25	31	25	25	44	44	17
10	1 1/4	31,75	32	38	31	31	54	54	21
—	1.260	32,00	—	—	32	—	—	—	—

NOTE: Inch sizes in parenthesis for "alternate sizes" D or E (under ¼ inch IC).

FIGURE 10-17

Industry-standard carbide identification system. (Courtesy of Kennametal Inc.)

IC		± tolerance on "IC"							
		class "M" tolerance						class "U" tolerance	
		shapes S, T, C, R & W		shape "D"		shape "V"		shapes S, T & C	
inch	metric	inch	mm	inch	mm	inch	mm	inch	mm
5/32	3,97			—	—	—	—	—	—
3/16	4,76			—	—	—	—		
7/32	5,56	.002	0,05	.002	0,05	.002	0,05	.003	0,06
1/4	6,35								
5/16	7,94								
3/8	9,52								
7/16	11,11	.003	0,06	.003	0,06	.003	0,06	.005	0,13
1/2	12,70								
9/16	14,29								
5/8	15,88	.004	0,10	.004	0,10	.004	0,10	.007	0,18
11/16	17,46								
3/4	19,05								
7/8	22,22	.005	0,13	—	—	—	—		
1	25,40			—	—	—	—	.010	0,25
1 1/4	22,22	.006	0,15	—	—	—	—		

IC		± tolerance on "B"							
		class "M" tolerance						class "U" tolerance	
		shapes S, T, C, R & W		shape "D"		shape "V"		shapes S, T & C	
inch	metric	inch	mm	inch	mm	inch	mm	inch	mm
5/32	3,97			—	—	—	—	—	—
3/16	4,76			—	—	—	—		
7/32	5,56	.003	0,06			—	—		
1/4	6,35			.004	0,11	—	—	.005	0,13
5/16	7,94					—	—		
3/8	9,52					.007	0,18		
7/16	11,11					—	—		
1/2	12,70	.005	0,13	.006	0,15	.010	0,25	.008	0,20
9/16	14,29					—	—		
5/8	15,88					—	—		
11/16	17,46	.006	0,15	.007	0,18	—	—	.011	0,27
3/4	19,05					—	—		
7/8	22,22			—	—	—	—		
1	25,40	.007	0,18	—	—	—	—	.015	0,38
1 1/4	22,22	.008	0,20	—	—	—	—		

—	4	3	2	□	□	□
	12	04	08	□	□	□
	[5]	[6]	[7]	[8]	[9]	[10]

[6] thickness

thickness		symbol	
in	mm	inch	metric
1/32	0,79	0.5 (1)	—
1/16	1,59	1 (2)	01
5/64	1,98	1.2	T1
3/32	2,38	1.5 (3)	02
1/8	3,18	2	03
5/32	3,97	2.5	T3
3/16	4,76	3	04
7/32	5,56	3.5	05
1/4	6,35	4	06
5/16	7,94	5	07
3/8	9,52	6	09
7/16	11,11	7	11
1/2	12,70	8	12

NOTE: Inch sizes in parenthesis for "alternate sizes" D or E (under 1/4 inch IC).

hand of insert (optional) [8]

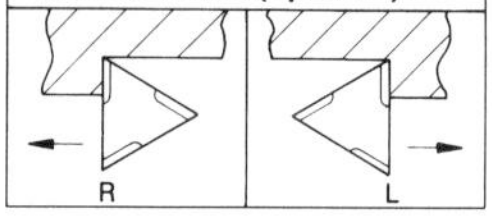

R   L

[7] corner radius

corner radius		symbol	
in	mm	inch	metric
.004	0,1	0	01
.008	0,2	0.5	02
1/64	0,4	1	04
1/32	0,8	2	08
3/64	1,2	3	12
1/16	1,6	4	16
5/64	2,0	5	20
3/32	2,4	6	24
7/64	2,8	7	28
1/8	3,2	8	32
round insert (inch)	—		00
round insert (metric)	—		M0

[9] & [10] cutting edge condition or chip control features (optional)

T — negative land

K — light feed chip control, double sided Kenloc insert

M — heavy feed chip control, deep floor Kenloc

N — narrow land Kentrol insert with chip control on one side

W — heavy-duty chip control, wide land Kenloc insert one side

J — polished to 4-microinch AA (rake face only)

UF — ultra-fine finishing

See Technical Section for additional conditions and chip control features.

FIGURE 10-17
Continued

Some carbide insert tools such as combination drills, similar to that shown in figure 10-18, can be used for facing and turning operations as well as for drilling a hole from solid on a CNC lathe—all with the same tool! In addition, cutting speeds and feeds are greatly increased, thereby improving and maximizing productivity levels. The high-technology indexable insert drilling bars (see figure 10-19) were designed to replace conventional twist drills and spade drills. These drills are capable of running up to 10 times faster, depending upon material type, because they run at coated carbide speeds and feeds. The indexable insert drilling bar uses multiple-edge, two-sided sided indexable inserts in most cases. Use of these indexable insert drills reduces cycle stop time because only the inserts need to be replaced, not the entire drill. The inserts can be indexed while still at the machine, and the use of inserts eliminates the need for tool resharpening.

Most carbide insert drills have other features that make them extremely advantageous over conventional twist drills. Indexable insert drills provide increased web thickness. This gives them the strength to handle high penetration rates. The larger shank diameters provide added rigidity and help avoid chatter. In drilling where a finish bore operation is required, a hole can be drilled very close to the final size desired; consequently, subsequent semifinish boring operations can be eliminated.

These types of drills, besides adding rigidity, operate at substantially higher metal removal rates than conventional high-speed twist drills or spade drills. For this reason, good machining practices are mandatory. Flying chips, for example, create a danger to the operator; a safety shield is usually required. Coolant is also necessary to cool the cutting edge and to backflush chips. Sufficient coolant must be continuously applied on these tools because of the higher chip removal rates, speeds, and feeds. In addition, drilling with carbide insert drills develops high thrusts. If the setup is not rigid, the forces will create chatter and side loading. This can result in broken inserts and damage to the drilling bar.

Carbide insert tools are constantly being researched and tested in modern manufacturing applications. As the quest for better and more efficient methods of production speeds up, so will the search for better and more efficient cutting tools.

Heat is the primary cause of insert wear and the destroyer of insert life. The shearing action of the metal being formed into a chip (see figure 10-20) and the rubbing of the chip against the cutting tool are the source of the heat. Other sources of insert wear include excessive or inadequate feeds, speeds, and depth of cut, as well as various forms of tool deformation.

The most common forms of insert deformation that cause wear are cratering and chipping. Cratering is generally caused by improper insert selection based on the alloy content of tungsten carbide with titanium and tantulum carbides. An improper alloy content causes a lack of resistance to crater and flank wear, resulting in cutting-edge deterioration and breakdown. Inserts designed for high-speed applications should contain large amounts of titanium to resist cratering.

FIGURE 10-18
Carbide insert
drill-face-turn tool.
(Courtesy of The
Valeron Corporation)

FIGURE 10-19
High-technology carbide insert
drill. (Courtesy of The Valeron
Corporation)

FIGURE 10-20
A turning operation
depicting heat as the
primary destroyer of
insert life. (Courtesy of
Cincinnati Milacron Inc.)

Chipping is caused by impact to the insert cutting edge or by vibration and lack of rigidity in the machining setup. Using an insert cutting edge that is not crack or vibration resistant or using too hard a grade of carbide for a particular application promotes chipping. Material may weld itself to the insert cutting edge if cutting speeds are too low. When the built-up material is removed, a portion of the cutting edge may also be removed, thereby causing a small chip in the insert.

Coated carbides and ceramics today permit impressively higher machining speeds and feeds than ever thought possible. These recent insert innovations and the wide variety of insert options and designs, coupled with application complexities, can make insert choice and selection a difficult, time-consuming, and costly undertaking. Consult competent tool and insert vendors when in doubt concerning insert type, style, and usage.

Many N/C and CNC machining applications require the use of cutting fluids. Coolant use is an important factor related to cutting tool life because of the following reasons:

FIGURE 10-21
A continuous flow of cutting fluid
over the cutting edge of the tool
is important to cutting tool life.
(Courtesy of Cincinnati Milacron
Inc.)

- It cools and carries heat from the cutting edge of the tool and workpiece. It is advisable to use a continuous flow of coolant over the cutting edge and entire operation (see figure 10-21) or else not to use it at all. Intermittent coolant use may cause insert thermal cracks.
- It lubricates and reduces heat friction at the face of the tool and allows smooth chip flow off the cutting edge. This reduces tool wear and power consumption while improving surface finish.

Cutting fluids have a secondary purpose of "damping down" flying dust or fine particles that are a safety hazard to the health of the operator.

Chip control, often regarded as a metal working nuisance, is generally handled in a hit-or-miss fashion. However, chip formation and flow (see figure 10-22) are generally functions of proper versus improper tooling, feeds, speeds, and depth of cut.

Chip control in heavy turning center stock-removal applications usually presents the greatest challenge. The high cutting forces and extreme heat will destroy all but the most rugged chip control devices. If the chip breaker is too narrow, for example, tight flying chips will occur and the corrective action is to widen the chip breaker. If the chip is too loose and fails to break or coil, the corrective action is to narrow the width of the chip. The breaker must be increased if the feed is increased in order to maintain consistent chip control performance. If the feed is decreased, chip breaker width should be decreased accordingly.

FIGURE 10-22
A heavy turning center stock-removal application showing chip formation and flow. (Courtesy of Cincinnati Milacron Inc.)

Actions similar to those described above help to better control chip formation and flow problems. In many cases, continued experimentation is inadequate, and the advice of professional tooling engineers should be sought.

Chips obtained from machining parts have value and are therefore salvaged. Chips fall down and away from the cutting action onto a continuously running conveyor system mounted either as part of or underneath the machine. They are then transported to large bins or chip containers, where they are dumped, compacted in some cases, and then sold.

TOOLHOLDERS

Any discussion of cutting tool considerations should include toolholders. Spindle-type toolholders (see figure 10-23) essentially hold and drive the entire assembly, which is made up of the basic cutting tool, adapters or collets, and the toolholder itself. Most toolholders conform to the ANSI industry standard for tapered V-flange toolholders. Dimensions are critical on flange, shank, and retention studs. Conformance to such rigorous tolerances and ANSI standards is mainly for safety and interchangeability requirements. The machine tool accepts only one basic size; #50 is commonly used on horizontal machining centers that have from 10 to 40 HP. The most important attribute of toolholders is that they be free from nicks, gouges, dirt, grit, and any other visible signs of damage. Lack of toolholder cleanliness can cause the entire tool assembly to "run out," resulting in part inaccuracies. Toolholders must be cleaned, inspected, reconditioned

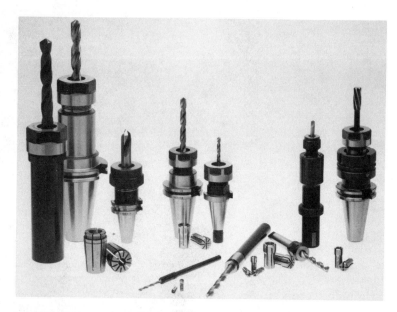

FIGURE 10-23
Spindle-type toolholders hold and drive the entire tool assembly.
(Courtesy of Kennametal Inc.)

(if possible), or scrapped if accuracy or safety becomes questionable. Toolholders must accurately drive the cutting tools and run true. In addition, machine tool spindles and tool matrix pockets should periodically be inspected and wiped out (while the machine is stopped and in the manual mode) to remove dirt, chips, and grit. Gouging of the machine spindle or the toolholder can result when pull-back clamping pressure is applied. Accepting anything short of an optimal tool assembly is settling for second-rate performance of the CNC machine tool.

Turning center toolholders that use disposable inserts, as shown in figure 10-24, permit indexing a carbide insert that has six to eight cutting edges in less than one or two minutes. If part tolerances are relatively open, the original tool setting can be maintained, and the machine is ready for production with no additional tool adjustment. Generally, as long as the same holder is used on an operation, the tolerances within the holder will have little effect on the ability of the insert to index to close tolerances.

Toolholder identification systems, used to identify tools with associated tool assembly numbers and other dimensional data, are available and are used in stand-alone, cell, and system applications. Automated identification systems are important because they are reliable, they save time, and they reduce human error. The most common of these systems involve the following:

- Bar code scanning
- Machine vision
- Radio frequency identification

FIGURE 10-24
Turning center toolholders that use disposable inserts permit indexing of
carbide inserts. (Courtesy of Kennametal Inc.)

- Optical character recognition
- The microchip

Two of the most widely used tool identification systems are bar code scanning
and the microchip. Bar codes are made up of binary digits arranged so that
the bars and spaces in different configurations represent numbers, letters, or
other symbols. In some cases, bar coded plates are attached to the outside of
the toolholder itself. Scanners that read bar codes (see figure 10-25) contain a
source of intense light produced by a laser or light-emitting diode aimed at the
pattern of black bars and spaces of varying widths. However, affixing bar codes
to the outside of the toolholder has in most cases proven ineffective because
they eventually fall off or get too dirty and greasy, distorting the scanner's ability
to read the bar code accurately.

A much more dependable identification system utilizes a microchip. Microchip identification entails the use of a microchip embedded in a sealed capsule that can be inserted in the toolholder. This system (see figure 10-26) uses a
noncontact read-only head that can be attached to toolchangers, fixtures, or tool
grippers. Reading is done with a proximity sensor and can be done at a distance
of up to 0.080 in.; the read time is less than 50 milliseconds, with an allowable
0.120-in. misalignment. The microchip can also be programmed off-line with
the required tool data.

FIGURE 10-25
Scanners that read bar codes
contain a source of intense
light produced by a laser or
light-emitting diode aimed at
a pattern of black bars and
spaces. (Courtesy of Kearney
and Trecker Corp.)

FIGURE 10-26
Microchip identification employs the use
of a microchip embedded in sealed capsule
that can be inserted in the toolholder and
read by a noncontact read-only head.
(Courtesy of Kennametal Inc.)

FIXTURING

Proper fixturing is also extremely important to successful N/C and CNC machining. A poorly designed or manufactured fixture often causes problems at the machine, holds up production, or produces much waste. If proper design considerations are applied, these costly and time-consuming problems can be avoided.

The basic function of a typical fixture (see figure 10-27) is to locate and secure the part for succeeding machining operations. This involves initial setup time for the workpiece to be loaded and clamped in the fixture for machining. Loading

FIGURE 10-27
A typical N/C machining center fixture. (Courtesy of Cincinnati Milacron Inc.)

and unloading constitute an important part of the nonproductive cycle time of each part. If this operation is simplified, more parts can be produced each hour. Therefore, fixtures should be designed to reduce part setup time.

N/C machines normally allow smaller batch lots, thereby reducing part inventories. Consequently, the fixture will be set up on the machine and used more frequently. This will warrant much consideration during design so as to reduce the setup time and simplify the process of locating and securing the fixture to the machine table.

The accuracy of all parts requiring special fixtures depends on the workholding device, regardless of how well the part is programmed or processed. Money spent taking the time to cover all aspects of locating and holding the workpiece accurately and securely will pay dividends when part and fixture reach the shop floor. This time and money should be spent in the initial design phase, before production begins.

The advantages of sound, economical design and accurate manufacture and assembly of fixtures are as follows:

- Reduced fixture-to-machine and part-to-fixture setup time
- Consistency of part accuracies
- Reduced errors and inaccuracies in part location
- Decreased cost per part
- Reduced inspection time
- Minimal fixture modifications or rework
- Faster and easier N/C data prove-out

Another important aspect of fixture design is that of clamping the workpiece. Clamps should always be placed as close to the support locations as possible. Placing unsupported clamps at any convenient location on the fixture could mean distortion of the part under clamping pressure. Clamp positions, for ease of loading and unloading, should also be given a priority. Clamps must avoid blocking hole locations and milling cuts that may interfere with part processing. Toolholders should be considered, particularly when the cutter is engaged in the workpiece, as long as they will not interfere with axis movements.

Clamping should become increasingly automated for larger part volumes. The types of clamps that should be used are pneumatic- and hydraulic-actuated clamps, toggle clamps, and cam-actuated clamps. Regardless of the type to be used or the number of parts to be processed, fixtures should always be designed with safety in mind. The operator should always be able to reach easily all clamps and any adjusting screws that may need attention. All sharp corners and projections should be minimized, and chip removal should be easy. The perfectly designed fixture is also the safest for those who must use it.

One overlooked aspect of fixturing that sometimes becomes evident after the fixture is in production is part orientation. Fixtures should always be designed to prevent incorrect loading and part orientation. Foolproofing of fixture design to prevent incorrect loading is critical; carelessness can occur when work is very repetitive. To foolproof the fixture design, there should be only one way the part can be located and clamped. This plan may take time to develop, but it will alleviate the problem of scrapped parts and broken tools.

Several ideas must be considered regarding the positioning of the part and fixture in relation to the machine table.

- *Will all the tools reach?* The maximum and minimum tool lengths should be considered to determine optimum positioning of the fixture-to-table and part-to-fixture relationships.

- *Is the fixture in a foolproof position relative to the machine table?* Much time and effort is often spent designing a foolproof part-to-fixture relationship, and the fixture is loaded and clamped 90 or 180° out of position. This can have expensive and hazardous consequences if the fixture is not marked to orient it with respect to the machine table.

- *Is the fixture designed and placed on the machine table to support efficient part processing?* Sometimes inefficient CNC moves are needed to avoid poorly designed and placed locating pads and clamp assemblies. If the machine has a rotary index table, the part-to-fixture and fixture-to-table relationships should ideally be positioned over the center of rotation. This should provide an equal distance from the center of rotation to each face to be machined. It should also promote programming simplicity in X-, Y-, and Z-axes.

Modular fixturing, which is gaining wide acceptance among manufacturers, employs reusable components that can be assembled and torn down based on

FIGURE 10-28
Assorted modular fixturing components. (Courtesy of QU-CO Modular
Fixturing, TE-CO Co.)

changing part requirements. Components (see figure 10-28) are manufactured
to close tolerances and can generally be assembled on a grid plate based on a
simple sketch. Sometimes referred to as "erector set" tooling, modular fixturing,
as shown in figures 10-29 and 10-30, affords users the following advantages:

- Fixtures can be assembled and torn down, and components can be reused.

- Dedicated fixture design, component manufacturing, inspection, and rework costs can be substantially reduced or eliminated.

- Elimination of fixture design and manufacturing saves valuable lead time
  and reduces coordination problems related to fixture design, manufacturing, and assembly.

- Hard fixtures (those dedicated to only one part or part family) are
  scrapped when part life becomes inactive; modular components can be
  salvaged and reused.

- If engineering change necessitates alteration, hard fixturing requires new
  lead time, rework, and inspection; modular components can easily be
  moved or replaced with other modular components as required.

- Hard or dedicated fixturing requires extensive storage facilities; modular
  components can be stored in bins and assembled as needed.

- The fixture designer is never locked into his or her design because alternative fixturing solutions can easily be tried by relocating clamps, stops,
  and locators.

- The part-to-machine setup and orientation time required for N/C jobs
  using table setups (conventional nondedicated work-holding components) can be reduced or eliminated; modular fixtures can be accurately
  re-created and assembled off-line in a fixture setup area in advance of
  machining operations.

FIGURE 10-29
An example of modular fixturing. (Courtesy of Cincinnati Milacron Inc.)

FIGURE 10-30
A modular fixture showing component buildup and part location. (Courtesy of Cincinnati Milacron Inc.)

Modular fixtures can be designed by trial and error, without a formal design, using a conventional fixture design approach. A major advantage of modular fixturing is that the fixtures can be constructed at the last minute, once the part to be machined is available. Many times, better fixtures are built by trial and error because the problems associated with part orientation can readily be visualized. It is not uncommon for new modular fixtures to be built completely from scratch in less time than it takes to prepare the cutting tools for the job. In addition, reconstructing a modular fixture that has been built once and had its design documented requires about a quarter of the time.

Companies with CAD systems can create a library of fixture components on their systems to further reduce design time, especially if the part to be fixtured is already in the CAD database. Once a CAD library of fixture components has been created on the system (or perhaps even purchased from the fixture component manufacturer), components never need to be re-created. A modular fixture designed on a CAD system is created by recalling the desired fixture components from the library and placing them where needed, until the fixture is complete. Utilizing the CAD fixture design methodology, a designer can try several different fixturing solutions via the CAD screen for the same part to determine the best design without actually having to build each fixture. Since CAD systems are inherently accurate, typically in excess of five decimal places, the fixture designer has the added advantage of being able to check for fixture component/workpiece interferences.

SOME SOUND TOOLING PRACTICES

Regardless of the N/C machine to be used, types of cutting tools employed, or method of holding and locating the workpiece, sound cutting guidelines and qualified tooling practices must be followed if a machine is to be used to its greatest capacity.

There is no easy analysis to be made for any machining operation. Material characteristics and a wide assortment of other conditions make it difficult to formulate general solutions to tooling problems. Each situation requires individual analysis after careful consideration of all the variables. There are, however, some general N/C tooling practices, reminders, and guidelines that warrant discussion.

1. *Check all tools* before they are used. This includes the cutting edges and tool body as well as holders, extensions, adapters, etc. They should all be in perfect order and able to act as a total tool assembly.

2. *Select the right tool* for the job, and use the tool correctly. Tools should be able to machine the workpiece to the desired accuracy. In purchasing cutting tools and toolholders, examine cost per part produced as well as the cost of the tooling package. In many cases, bargain tools cost more per part produced.

3. *Always choose the shortest drill length* that will permit drilling of the hole to the desired depth. The smaller the drill size, the smaller the allowable error; as drill size increases, the allowable error progressively increases.

4. *Check and maintain correct cutting feeds and speeds for all tools.* This idea can spell success or failure for any N/C installation. Optimum feeds and speeds may not always be achievable, depending on machinability and material characteristics, rigidity of setup, etc.

5. *Understand machine and control capabilities.* Additional tools are often purchased because built-in machine/control capabilities such as contouring are not used effectively. Most controllers sold are of the contouring type, but qualified personnel must still be able to recognize where and how their capabilities can be used to alleviate tooling problems.

6. *Care for tools properly.* This includes perishable tools, holders, drivers, collets, extensions, etc. They represent a sizable investment and should be adequately stored and reconditioned when necessary.

7. *Watch and listen for abnormal cutting tool performances.* Attention paid to actual metal removal processes can often prevent tool breakage problems, scrapped parts, and rework. Chatter and other vibration abnormalities can be corrected if detected early enough.

8. *Use correct tap drills.* Often the wrong drill is used for a certain size tap, and taps are sometimes broken as a result of negligence and incorrect tool selection.

9. *Holes to be tapped should be deep enough, free of chips, and lubricated prior to tapping.* Insufficient or incorrect lubrication can cause tap breakage, oversized threads, and poor surface finish.

10. *Use end mills as counterbores* where possible. They reduce inventory, produce a lower tool cost per piece, and produce a spotface or counterbore on a rough or angled surface more easily.

11. *The length-to-diameter ratio should not be exceeded on boring bars.* The boring bar length should never exceed four times the bar diameter. Failure to comply with this rule could result in chatter and inaccuracies in hole size.

12. *Select standard tools* whenever the operating conditions allow. Standard tools are less expensive, readily available, and interchangeable.

13. *Select the largest toolholder shank* the machine tool will allow. This will minimize deflection and reduce the tool overhang ratio.

14. *Select the strongest carbide insert* the workpiece will allow. This will increase overall productivity and lower the actual cost per insert cutting edge.

15. *Use negative rake insert geometry* whenever the workpiece or the machine tool will allow for it. This will double the cutting edges, provide greater strength to the insert, and dissipate the heat.

16. *Select the largest insert nose radius,* but the smallest insert size, that either the workpiece or machine tool will allow. The smallest size insert will be less expensive; using the largest insert nose radius will improve finish, dissipate heat, and provide greater strength.

17. *Select the largest depth of cut and the highest feed rate* that either the workpiece or machine tool will permit when carbide inserts are being used. This will improve overall productivity and have a negligible effect on tool life.

18. *Know the workpiece material and its hardness.* It is essential to have a thorough knowledge of the material's machining characteristics. If little is known, start at the lowest given cutting speed for that particular material and gradually increase the speed until optimum results are obtained.

19. *Select the cutting speed in relation to the physical properties of the workpiece.* The Brinell hardness number usually gives a good indication of the material's relative machinability.

20. *Increase cutter life by using lower cutting speeds and increasing the feed rate* to the limits allowed by the desired results, the setup rigidity, and the strength of the tool.

21. *For maximum cutter life, the feed should be as high as possible.* Doubling the feed (measured as chip per tooth) will double the stock removal per unit of time without appreciably decreasing cutter life.

22. *Use low cutting speeds for long cutter life.* However, soft, low-alloy materials can be machined at high cutting speeds without seriously affecting cutter life.

23. *Excessive cutting speeds generate excessive heat, resulting in shorter cutter life.* Chip and cutter tooth discoloration are good indications of excessive cutting speeds.

24. *Be aware that when the tool's cutting edges quickly become dull,* without chip or tooth discoloration, either the workpiece is very abrasive or there is a high resistance to chip separation. The cutting speed must be reduced.

25. *Use a sharp corner milling cutter only when the job calls for milling to a sharp corner.* Otherwise, use a cutter with a large corner chamfer. Greater stock removal with lower horsepower requirements will result.

26. *Climb milling will allow higher cutting speeds.* In addition, it will improve finish and lengthen cutter life.

27. *Coarse tooth end mills are preferable for roughing cuts.* Although some operators prefer fine tooth end mills for finishing cuts, it is possible to obtain finishes by increasing the cutting speed and decreasing the feed (chip load per tooth) of the coarse tooth end mill. Thus, the same end mill can be used for roughing and finishing.

28. *Direct cutting forces against the solid portion of both the machine and the fixture.* If work is held in a vise, direct the cutting forces against the solid jaw.

29. *Use coolants to get maximum cutter life* and to permit operation at higher cutting speeds. Although coolant is not normally used when milling cast iron, if a jet of air is applied as a coolant, finish can be improved and cutter life can be lengthened.

30. *Fixture design should be simple,* and standard components should be used whenever possible.

31. *The fixture and part should be located positively.* Rough, nonflat parts should be supported in three places and located on tooling holes if possible.

32. *Fixtures and parts should be readily accessible and movable* during any part of the machining operation and replaceable in exactly the same position.

33. *Fixture design should be simple and foolproof* for part loading to fixture and fixture loading to table.

34. *Always keep clamps close to fixture supports* in fixture design, and consider safety when designing for load and unload capabilities.

Following these tooling practices will help avoid potential tooling problems on N/C and CNC machines. Consideration should be given to these points for successful use of N/C and CNC machine tools.

REVIEW QUESTIONS

1. Explain, in your own words, why tooling considerations are so important to success on an N/C machine.

2. What is the most important criterion during selection of the common twist drill for an N/C machine?

3. Why is center, or spot drilling important to hole location?

4. When should spade drills be used instead of common twist drills? Why?

5. Discuss the advantages of multiple-diameter, multiple-land tools over single-diameter, single-land tools.

6. What are the most common causes of tap breakage on N/C machines? How can they be avoided?

7. What are hand taps? When should hand taps be used?

8. What is the primary difference between spiral pointed, or gun, taps and other taps? When should spiral pointed taps be used?

9. When should spiral-fluted taps be used? What are their primary advantages?

10. Explain how fluteless taps work. What is the most important point to remember when fluteless taps are being used?

11. Discuss the importance of reaming on an N/C machine. Why are floating toolholders usually not required for reaming operations on N/C machines?

12. What is meant by the length-to-diameter ratio (L/D) in boring? What is the general rule for one to follow when applying this ratio for boring bars?

13. Describe how end mills and a machining center's contouring capability might be used in place of additional boring bars.

14. When should end mills be used for counterboring? Describe the main advantages of using end mills rather than ordinary counterbores.

15. What are some advantages of high-technology indexable insert drilling bars?

16. Discuss other factors that must be considered when carbide insert tools are being used (for example, flying chips, coolant use, thrust, setup rigidity, etc.).

17. Why is fixturing so important to the success of an N/C machine?

18. What are the advantages of sound economical fixture design?

19. What is meant by foolproofing a part to fixture and fixture to machine table?

20. Why is it important to inspect thoroughly all cutting tools and fixtures before use?

21. What are the two most common forms of insert deformation and wear, and how are they caused?

22. What are the advantages of using cutting fluids in machining applications?

23. Name the two most widely used tool identification systems.

23. List four advantages of modular fixturing over hard fixturing.

CHAPTER 11

Advanced CNC Applications and Integration

OBJECTIVES

After studying this chapter, you will be able to:

- Identify the two types of computer-aided process planning (CAPP)
- Discuss advanced CAD, CAM, and CIM applications and their importance to productivity improvements and plant integration
- Describe manufacturing cells and systems and discuss their impact on future industrial capabilities
- Name some advanced programming and machining concepts

BEYOND THE PROCESSOR LANGUAGES

Although APT and other processor languages such as COMPACT II have been around for some time and continue to be used by some small and large companies, their use continues to give way to more of the graphics-based N/C systems. However, change comes slowly to many businesses. Even though significant technological advancements, ease of use, and overall system price increase the desirability of such systems, many companies have a multitude of legacy programs on APT- and derivative-based systems. This raises the question of whether to convert all active programs over to a graphics-based system or maintain two systems—the old one to make changes to existing parts already programmed and the new one to create new parts. Many companies continue to debate how to deal with this issue, desiring to further advance graphics-based systems but deeply entrenched in the legacy APT-based systems.

Changes have also occurred in the multitude of postprocessors written to support APT and other languages. Many postprocessors were written for hard-wired controls and did not contain some of the advanced routines, such as stored parametric subroutines and repeatable patterns, of modern CNC units. Postprocessors today have built-in capabilities for stored parametric subroutines. These sophisticated techniques reduce the need in some cases for large general processor languages by eliminating from the actual program some routines that may now be resident in the CNC unit. The recent developments in

CNC technology are primarily due to rapid microprocessor developments and enhancements. In addition, they further reduce computer application requirements by making shop floor programming justifiable and efficient.

COMPUTER-AIDED PROCESS PLANNING (CAPP)

Process planning involves creating detailed plans of the manufacturing steps and equipment necessary to produce a finished part. Workpiece requirements call for detailed analyses and accurate descriptions prior to the actual manufacturing process. A large assortment of machines and operations, as well as many different workers with a variety of skills, may be involved in the production of a specific part.

The computer lends itself well to the vital process-planning function with two different approaches. One is called the *variant*, or *similar part*, method of process planning, and the other is *generative*. Both will produce an equal or accurate process plan, but most computer applications are of the variant type because the software is easier to develop and new process plans are based on previous ones.

The variant method had its beginnings with the group technology concept, along with parts classification and coding systems. Group technology is a manufacturing philosophy based on the idea that similarities occur in the design and manufacture of component parts. These parts can be classified into groups, or families, if the basic configurations and attributes are identified. A reduction in expenses can be achieved through the structured classification and grouping of parts into families based upon engineering design and manufacturing similarities.

Using the variant method, CAPP groups families of parts by a structured classification and coding plan. All previously processed parts are coded via this method. The parts are then divided into part families, such as rotational and bar and rail, based on general configuration. A standard order of operations or sequences is stored on the computer for each part family. When a new part is ready for planning, the classification or group technology code for the new part is used to compare and retrieve the standard process plan for that part family. Editing capabilities further enable the process planner to alter the standard order of operations for final refinement. The completed process plan is then stored on the computer database by part number.

In generative process planning, the parts are again broken into part families, and a detailed analysis is made for each part family to determine individual part operations. This type of system develops the actual operation sequence based on the part geometry, usage requirements, material size and configuration, and available equipment. The generative approach creates process-planning logic for the part family groupings. The logic is then stored internally as a decision model. As new workpieces require process planning, analyses must be conducted to determine and compare the features incorporated in the decision model with those on the actual part. The family part decision model is then retrieved, and a routing sheet is generated by processing the decision model with the new workpiece attributes.

A generative system must be driven by much more elaborate and powerful software than a variant system. Development and optimization work continues on the variant and generative approaches to process planning. Both systems, however, build the needed decision-making logic and planning ability into the computer, rather than relying on a decreasing experience level in the process-planning work force.

ADVANCED CAD/CAM

Although existing PC, workstation, and mainframe-based systems provide a high level of user functionality and capability, additional developments and enhancements continue to take place, particularly in the areas of feature-based machining, generative N/C, and knowledge-based systems.

All parts in any manufacturing plant can be represented by a combination of a relatively limited number of assignable features (holes, slots, contours, threads, etc.). Features can be described by combinations of shape definition and geometrical and technological data. They may be implicit (standard features like holes and slots), or they may be explicit (a nonstandard, user-defined feature such as a concave pocket). The majority of implicit or standard features, although not exactly the same across products or across companies, are common across industry. Experience has shown that the number of features ranges from 20 for simple parts to roughly 150 for complex manufacturing tasks. In a typical manufacturing environment, about 25 features can be used to describe prismatic and flat parts and about 15 additional features for round parts.

Solid-model machining duplicates the manufacturing process by using true-volume models of a part (see figures 11-1* and 11-2*). It gives the user an unambiguous pictorial representation of the actual machining process as the tool motion simulation removes the material. Part features such as pockets, slots, holes, threads, and so on can be defined in the model and then machined as specific features. The key, then, is having developed, proven, and encapsulated machining logic and generative N/C modules for the various part features, as depicted in figure 11-3. Each feature-based machining logic module would then do the following:

- Divide a feature into machining steps (rough, semifinish, finish)
- Find a suitable cutter in the database to machine each step
- Calculate machining parameters such as feed, speed, and depth of cut
- Generate tool path direction and number of passes to machine the feature

Each feature-based machining module would also contain the following information:

- Raw material geometry
- Final feature geometry

---

*See also color insert this chapter.

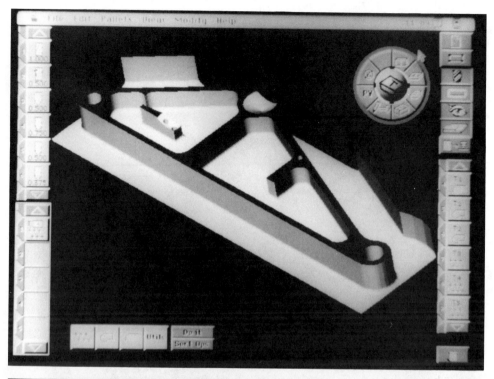

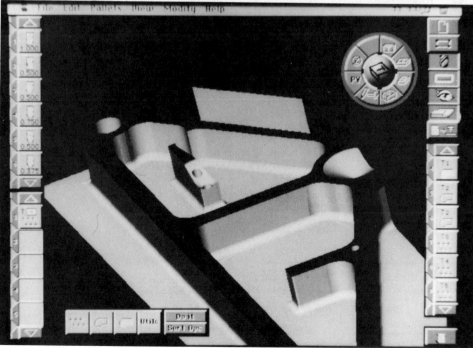

FIGURE 11-1
Graphics systems today actually allow users to view the three-dimensional part model and its simulated machining process, highlighting unfinished surfaces, excess stock, depths of cut that are too shallow or too deep, etc. (Courtesy of Gibbs and Associates)

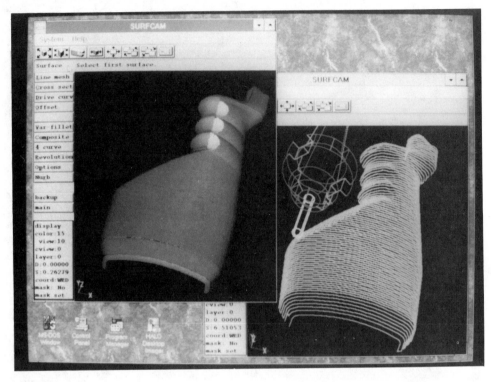

FIGURE 11-2
Design engineering master part models contain all the product definition data to be passed to manufacturing. (Courtesy of Surfware Inc.)

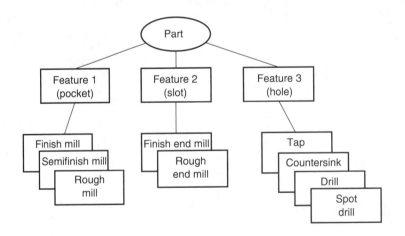

FIGURE 11-3
Proven and encapsulated feature-based machining logic modules can be defined in the part model and then machined as a specific feature.

- Geometry of the feature's immediate environment
- Type of material to be machined
- Required tolerances
- Available machine tools
- Cutting tools available for the designated machine

Design attribute information would already be available via the master model and such engineering specification data as required tolerances, final feature geometry, and material type. Users can then input remaining feature variables and parameters and override conventional machinability data where required. But the resident and preprogrammed generative N/C machining modules can take care of most of the machining for any standard part feature.

The developing software directions for solid modeling, generative N/C, and knowledge-based machining are being driven by, but are not limited to, the need to do the following:

1. Capture and retain retiring worker's manufacturing knowledge and experience
2. Compress the entire "design through manufacturing to assembly" cycle and bring products to market faster
3. Add consistency and standardization to the manufacturing process, thereby reducing variation and improving quality
4. Meet changing customer demands and expectations for reduced cost, improved quality, and faster product delivery
5. Meet the demands of increasing and intensifying global competition

As the sophisticated capabilities of solid-model machining, generative N/C, and knowledge-based systems continue to develop and mature, they will further automate the N/C programming task, provide for increased manufacturing predictability and dependability, and permit a reduced level of machining expertise in programming personnel. Machining knowledge and expertise will be captured in the preprogrammed, feature-based machining modules, enabling faster "up-front" producibility and cost determination and shortening the product's overall time to market.

In order that full advantage may be taken of these available software applications, their integration must occur and become seamless. Users need appropriate hardware and the ability to quickly access engineering and manufacturing data through simple software window-like user interfaces similar to those of the Apple Macintosh. As advanced CAD/CAM and other manufacturing applications further develop, their integration will surely yield tremendous productivity gains in the years ahead.

COMPUTER INTEGRATED MANUFACTURING (CIM)

What is CIM? Simply stated, CIM is a systems approach toward manufacturing. CIM attempts to integrate the common aspects of business and manufacturing so that information can be shared in a timely manner without duplication or errors caused by duplication. It is the integration of the design, manufacturing, and business functions of an organization through the use of computers and commonly shared, easily accessible data, for the purpose of improving company effectiveness. CIM incorporates company business strategies and includes the generation and collection of data as well as the dispersal and use of data through the involvement of several technologies. CIM includes the integration of the business information systems with the factory floor systems—in other words, the integration of manufacturing information systems with process control. CIM is part of a strategy to ensure the long-term survivability and profitability of a business. It is also the strategy by which companies can organize the various hardware and software components of their computer systems.

CIM is a journey, not a destination. It is not an off-the-shelf product that can be bought or developed independently of the business in which it is to be applied. In addition, each CIM implementation is as unique as each business. It implies more than just getting the various pieces of computer hardware in the business to communicate with each other, because it is as much an attitude and a state of mind as anything else. To be successful, CIM must start with strategies developed for the business. These business strategies must then drive the development of the manufacturing strategies. Then, the manufacturing strategies should drive the development of the CIM strategies, which should finally close the loop with the CIM strategies capable of achieving the original business strategies.

To understand the overall goals of CIM, its specific integration goals, and the goals for specific systems, it is necessary to understand CIM's architecture. It is typically divided into five levels, from the factory floor at the lowest level, to the corporate level at the top. This general CIM architecture is depicted in table 11-1.

TABLE 11-1 A General CIM Architecture

Level	Functions
Level 4: Corporate level (mainframe computers)	Centralized, common, accessible data (financial, purchasing, personnel, etc.)
Level 3: Plant-wide level (minicomputers or mainframes)	Production scheduling/inventory control Resource management Coordination with lower levels
Level 2: Department level (microcomputers)	Real-time data acquisition Real-time supervisory control of work cells
Level 1: Machine control/work cell level (microprocessors, PLCs, DNC systems, CNCs)	Real-time control of manufacturing activities
Level 0: Factory floor level	Performing of manufacturing activities

Without an integration strategy through CIM, there is little improvement in total company effectiveness. For companies to be truly successful with CIM, continued information integration and information sharing must be the essence of the strategy. CIM must be viewed as a long-term, evolving strategy—a continuing, ongoing program of integration and advancement.

MANUFACTURING CELLS AND SYSTEMS

The definition of a manufacturing cell in its broadest sense implies the logical arrangement, or clustering, of stand-alone manual or N/C equipment into groups of machines so as to process parts by part family. Additionally, by definition, processing parts in a manufacturing cell includes completing as much of the workpiece processing as possible within the cell before moving it to the next sequential processing, stocking, inspection, or assembly station. Today, the term *manufacturing cell* is much broader and further implies some level of automated part loading, unloading, delivery, or exchange to the clustered machines; and yet the term can take on a variety of different meanings, depending on a manufacturer's application, state of technological development and understanding, and particular point of view.

Basically, manufacturing cells can be divided into four general categories:

1. Traditional stand-alone N/C machine tool
2. Single N/C machine cell, or minicell
3. Integrated multimachine cell
4. FMS

The stand-alone N/C or CNC machine tool (see figure 11-4) is characterized as a limited-storage, automatic tool changer and is traditionally operated on a 1:1

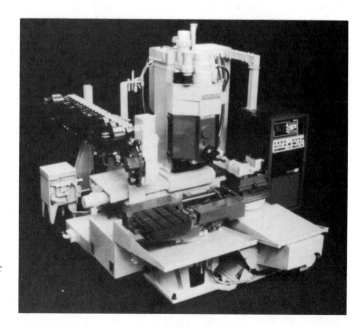

FIGURE 11-4
The traditional stand-alone N/C machine tool is typically operated on a 1:1 machine-to-operator ratio. (Courtesy of Cincinnati Milacron Inc.)

FIGURE 11-5
Some stand-alone N/C machine tools have been grouped together in a conventional part family cell arrangement, but still operate on a 1:1 machine-to-operator ratio. (Courtesy of Cincinnati Milacron Inc.)

machine-to-operator ratio. It has been the mainstay of capital acquisitions in the metal-cutting industry since the mid-1960s and continues to generate the bulk of CNC machine tool sales.

In many cases, stand-alone N/C machine tools have been grouped together in a conventional part family manufacturing cell arrangement (see figure 11-5), but still operate on a 1:1 machine-to-operator ratio. Machines within a cell of this type have sometimes been painted a similar color to further add cell distinction for a particular group of parts and to differentiate the group from other cells (for example, red cell, green cell, blue cell).

Some stand-alone N/C machines are characterized and operated as a cell by virtue of a change in the machine-to-operator ratio. These machines are usually redundant and are operated on a 2:1 or in some cases 3:1 machine-to-operator ratio (see figure 11-6). Because one operator is running more than one machine, the group of machines that the one operator is therefore responsible for is sometimes referred to as a cell. Part cycle start and stop times in a cell of this type are controlled and staggered by the operator so that one machine is always running while the other is idled for loading or unloading.

The single N/C machine cell (see figure 11-7) is characterized by an automatic work changer with permanently assigned work pallets or a conveyor-robot arm system mounted to the front of the machine, plus the availability of bulk tool storage. There are many machines with a variety of options, such as automatic probing, broken-tool detection, and high-pressure coolant control, that fall into this category. The addition of these and other special option features enables many single N/C machines to operate as self-contained cells. The single N/C machine cell, or minicell, is rapidly gaining in popularity, functionality, and affordability because it can be purchased for a fraction of the cost of a complete flexible manufacturing system (FMS) and can be programmed and loaded with parts to run unattended for several hours. Unattended operation of a single

FIGURE 11-6
Some N/C machines are grouped as a cell and operate on a 2:1 or 3:1 machine-to-operator ratio. (Courtesy of Cincinnati Milacron Inc.)

machine cell affords users the opportunity to gain some of the advantages of an FMS, such as increased spindle utilization and reduced direct labor, while increasing their knowledge and confidence about unattended machining and automated manufacturing capabilities.

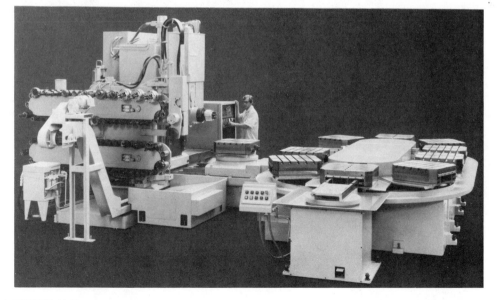

FIGURE 11-7
Single N/C machine cell with a permanently assigned and attached pallet work-holding system. (Courtesy of Cincinnati Milacron Inc.)

The integrated multimachine cell is made up of a multiplicity of metal-cutting machine tools, typically all of the same type, which have a queue of parts, either at the entry of the cell or in front of each machine. Typically, multimachine cells are either serviced by a material-handling robot (see figure 11-8), or parts are palletized in a two- or three-machine, in-line system (see figure 11-9). The typical application of a multimachine cell serviced by a robot (see figure 11-10) is high-volume production of a small, well-defined, design-stable family of parts. Machines can be different in a cell of this type, and workpieces can be progressively moved, for example, from a turning center to a center-type grinder for part completion. Palletized in-line cells can also be applied to either high or low variety and volume production applications. Material handling links together a group of flexible general-purpose machine tools utilizing common pallet design with prefixtured parts on pallets.

The flexible manufacturing system (FMS), sometimes referred to as a flexible manufacturing cell (FMC), is characterized by multiple machines, automated random movement of palletized parts to and from processing stations, and central computer control with sophisticated command-driven software. The distinguishing characteristics of this cell are central computer control; the automated

FIGURE 11-8

A multimachine cell may be serviced by a material-handling robot. (Courtesy of Cincinnati Milacron Inc.)

FIGURE 11-9
An in-line, dedicated manufacturing cell. (Courtesy of Cincinnati Milacron Inc.)

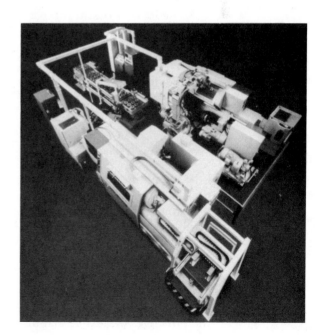

FIGURE 11-10
Typical application of a multimachine cell: a material-handling robot progressively moving a part from machine to machine. (Courtesy of Cincinnati Milacron Inc.)

flow of raw material to the cell; complete machining of the part; part washing, drying, and inspection within the cell; and removal of the finished part.

An FMS is made up of hardware and software elements. Hardware elements are visible and tangible, such as CNC machine tools, queuing carousels (part parking lots), material-handling equipment (robots or automatic guided vehicles), central chip removal and coolant systems, tooling systems, coordinate measuring machines (CMMs), part cleaning stations, and computer hardware equipment. Software elements are invisible and intangible, such as N/C programs, traffic management software, tooling information, CMM programs, workorder files, and FMS software. A typical FMS layout is shown in figure 11-11.

A true FMS, as shown in figure 11-12, can handle a wide variety of dissimilar parts, producing them one at a time in any order as needed. (Very few so-called FMSs meet this strict definition.) To operate efficiently in this mode, an FMS must have several types of flexibility. It needs the flexibility to adapt to varying volume requirements and changing part mixes, to accept new parts, and to accommodate design and engineering alterations. FMS also requires the flexibility to cope with unforeseen disturbances such as machine downtime or last-minute schedule changes, and to grow and change with the times through system expansion and configuration modification. These types of flexibility are made possible through computers and appropriate FMS software.

The following are advantages of an FMS:

- Increased machine utilization (spindle running time)
- Reduced work in process
- Increased part throughput
- Improved quality
- Reduced inventory
- Reduced personnel
- Accurate scheduling
- Reduced lead time

Many indirect benefits of FMS are important but difficult to quantify. Round-the-clock operations mean faster response time, more efficient assimilation of design alterations, and greater responsiveness to changes in the marketplace. Furthermore, within an FMS there is greater process control. This leads to improved consistency of quality, better predictive abilities, and less scrap and rework.

Almost all low- and mid-volume manufacturers who face worldwide competition will need some increased form of flexible manufacturing to improve efficiency and effectivity on the shop floor. Ultimately, they will want to apply the principles of flexible manufacturing throughout their operations.

KEY

1. Four CNC Machining Centers

2. Four tool interchange stations, one per machine, for tool storage chain delivery via computer-controlled cart

3. Cart maintenance station. Coolant monitoring and maintenance area.

4. Parts wash station, automatic handling

5. Automatic Workchanger (10 pallets) for online pallet queue (parking)

6. One inspection module — horizontal type coordinate measuring machine

7. Three queue stations for tool delivery chains

8. Tool delivery chain load/unload station

9. Four part load/unload stations

10. Pallet/fixture build station

11. Control center, computer room (elevated)

12. Centralized chip/coolant collection/recovery system (----- flume path)

13. Three computer-controlled carts, with wire-guided path — AGVs (Automatic Guided Vehicles)

↻ Cart turnaround station (Up to 360° around its own axis)

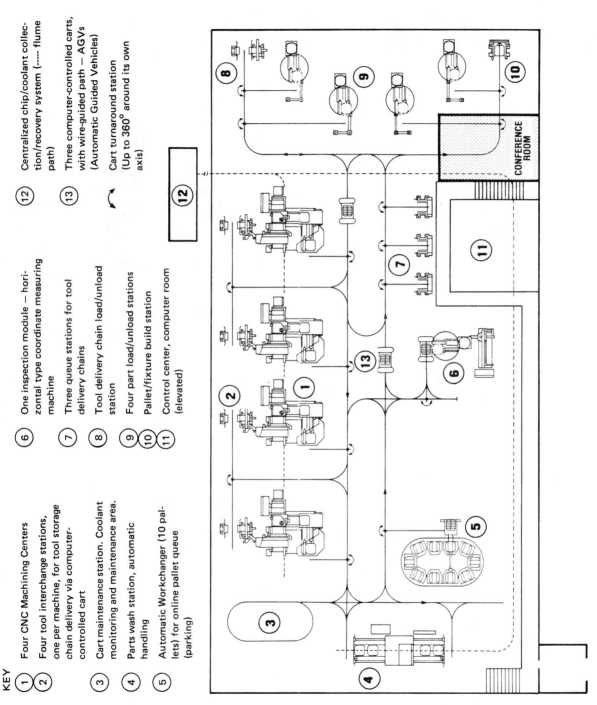

FIGURE 11-11

A typical FMS layout, with principal components identified. (Courtesy of Cincinnati Milacron Inc.)

FIGURE 11-12
A full-scale, installed, and operational flexible manufacturing system. (Courtesy of Cincinnati Milacron Inc.)

LOOKING AHEAD

What began as an idea created in the early 1950s by John Parsons has developed into a manufacturing concept that has revolutionized the metalworking industry. Numerical control has advanced from a stand-alone method of its own to a subset of a larger and much broader CAD/CAM, CIM, and advanced manufacturing industry.

Machining operations that use robot applications (with no human operator), similar to that shown in figure 11-13, will continue to play a significant role in the future of manufacturing and numerical control because of the following reasons:

- A declining percentage of the U.S. work force choosing careers in manufacturing
- Increased production pressure from foreign competition
- Internal inefficiencies that create costly in-process inventory and tie up working capital

FIGURE 11-13
Automated material-handling systems such as this robotic application will continue to be used for unattended machining operations. (Courtesy of Cincinnati Milacron Inc.)

- Increased availability of automated systems for part loading and unloading
- Significant advancements in electronics and microprocessor technology
- The need for predictable and dependable productivity levels

Advanced applications of CNC continue to expand into areas other than the more traditional metal cutting and forming. Composite manufacturing is one area being developed through enhanced machine and CNC technology.

The aerospace industry has been the prime initiator of composite manufacturing, mainly because of its need for the high strength-to-weight characteristics of composite materials. Kevlar, graphite, or boron fibers combined with a bonding matrix, such as epoxy, provide lightweight structures that have strength and flextural properties exceeding those of aluminum. The "laying up," or forming, of composite parts is a rigorous process of building up successive layers of resin-impregnated material to construct the general part shape and thickness. Application of pressure and heat, generally in an autoclave, then cures the fiber resin buildup to a finished rigid state.

Machines employing the use of this advanced technology are multiaxis, tape-laying, gantry machines similar to that depicted in figure 11-14. The tape-laying head (see figure 11-15) can precisely lay ply-on-ply layers of very thin

FIGURE 11-14
A CNC, multiaxis, gantry, composite, tape-laying machine. (Courtesy of Cincinnati Milacron Inc.)

FIGURE 11-15
A CNC, composite, tape-laying head that can precisely lay ply-on-ply layers of very thin tape at speeds up to 100 FPM. (Courtesy of Cincinnati Milacron Inc.)

tape (0.005 to 0.014 in. thick) at speeds up to 100 FPM. This includes automatic dispensing of the tape from a large-diameter supply roll, placing it directly on the underlying mold, applying the proper compaction pressure to seat and debulk, taking up and storing the backing paper, and at the end of each course cutting the tape (but not the backing paper) to the required length and angle.

Another recent and important CNC trend is the development of CNC waterjet and abrasive waterjet machining. Today, there is no American airplane, car, or house that does not have something in it that was cut with a waterjet or an abrasive waterjet.

A waterjet or abrasive waterjet (see figure 11-16) is a single-point sandblaster, so to speak, which may be pointed in almost any direction, does not wear out, is very safe to use, and causes very little (if any) damage to the material being cut. The process produces no thermal or mechanical distortions. After the material is cut, the water or water-abrasive stream is collected in a tank or reservoir.

Waterjet machining can cut a wide variety of metal and nonmetal materials by using a small-diameter (0.004 to 0.024 in.) focused stream of water passing through an orifice under high pressure (20,000 to 60,000 psi) traveling at high

FIGURE 11-16
A CNC abrasive waterjet machine and its material removal process. (Courtesy of Jet-Edge Corporation)

velocity (1,700 to 3,000 ft./sec.). When the force of the waterjet impacting the surface exceeds the strength of the material, the material is cut.

Materials can be cut multidirectionally, without ragged edges (unless the traverse speed is too high) and without heat, and generally more quickly than with a bandsaw. There are no thermal delamination or deformation problems when a waterjet is properly applied, and dust is nonexistent. Waterjet and abrasive waterjets handle materials that gum up saws or suffer unacceptable heat damage from other processes. Plasma and lasers, for example, leave a heat-affected zone on the material they cut.

Sometimes referred to as hydrodynamic machining, waterjet machining has been further improved by the addition of abrasives such as silica and garnet into the stream for use in cutting metals, composites, and other hard materials. Waterjet and abrasive waterjet machining continues to evolve, and the application has a rapidly growing acceptance rate among manufacturers.

Continued improvements will also take place in the area of machine tool controllers. CNC units will continue to be enhanced, utilizing high-level user graphics, more powerful microprocessors, additional memory, and advanced communication systems. These new systems will be capable of accessing engineering data directly from an engineering database whether in the same building or thousands of miles away, downloading engineering geometry directly to the MCU for part processing and bypassing off-line programming entirely.

Part programming will become increasingly automated and centered around feature-based technology. As discussed earlier in this chapter, feature-based technology depends on establishing and cataloging the representative features that make up the parts under consideration. Features are not exactly the same across products or across companies, but the majority of the features are common across industry.

Systems using feature-based decision logic are already in use in one form or another in several companies. Decision logic will also be provided for expert-intensive interpretation of engineering and manufacturing data, thereby greatly minimizing time-consuming, error-prone, and repetitive human decision making. The evolution and development efforts of these kinds of systems continue as researchers and manufacturers continue to find better and faster ways to bring products to market.

CAD systems also provide powerful programs such as finite element analysis to test and predict patterns of stress and strength in addition to calculating volume and weight.

(Courtesy of Surfware, Inc.)

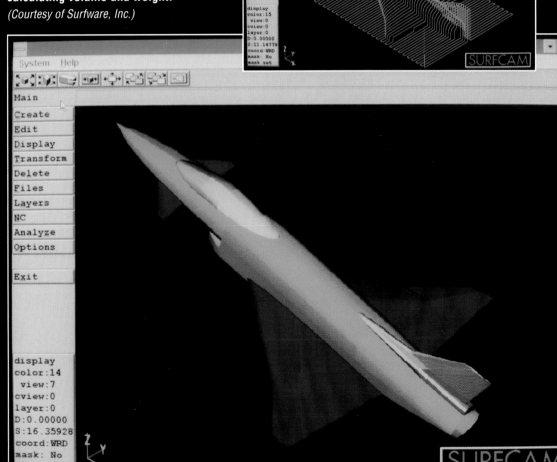

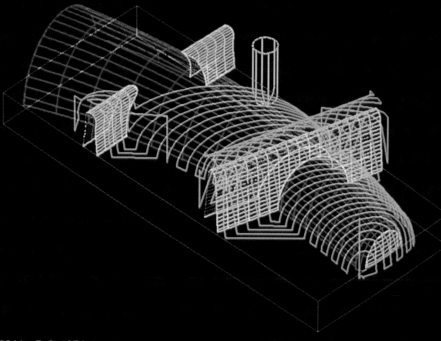

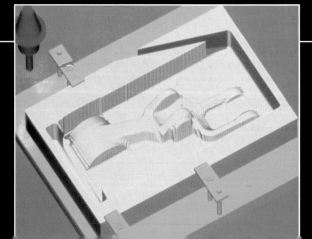

Some graphics-based systems consider the complete cutter and holder geometry in relationship to the workpiece to generate gouge-free tool paths. *(Above, Courtesy of CNC Software, Inc; Right, Courtesy of CGTech, Inc.)*

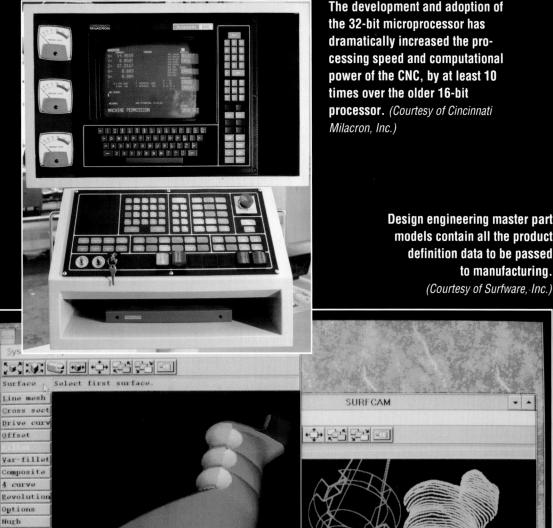

The development and adoption of the 32-bit microprocessor has dramatically increased the processing speed and computational power of the CNC, by at least 10 times over the older 16-bit processor. *(Courtesy of Cincinnati Milacron, Inc.)*

Design engineering master part models contain all the product definition data to be passed to manufacturing. *(Courtesy of Surfware, Inc.)*

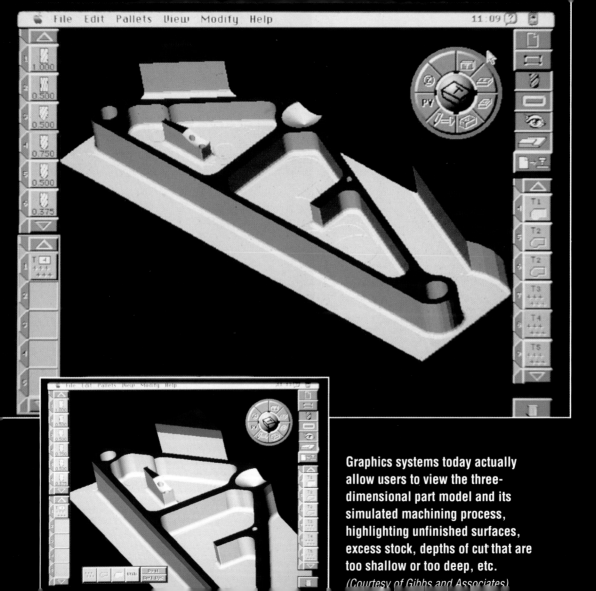

Graphics systems today actually allow users to view the three-dimensional part model and its simulated machining process, highlighting unfinished surfaces, excess stock, depths of cut that are too shallow or too deep, etc.
(Courtesy of Gibbs and Associates)

REVIEW QUESTIONS _____

1. What does CAPP mean? What is its overall impact on manufacturing?
2. Explain the difference between variant and generative process planning.
3. Describe some advanced CAD/CAM applications.
4. What is meant by a part *feature?*
5. What information is required in each feature-based machining module?
6. What forces are driving the need to develop solid modeling, generative N/C, and knowledge-based systems?
7. Name and describe the four general categories of manufacturing cells.
8. List four reasons why unmanned machining operations should continue to gain acceptance.
9. Describe what makes a manufacturing system "flexible."
10. List five advantages of FMS.
11. Describe feature-based manufacturing, and briefly discuss its potential.

APPENDIX A _____

EIA AND AIA NATIONAL CODES

Preparatory Functions

G word	Explanation
G00	Denotes a rapid traverse rate for point-to-point positioning.
G01	Describes linear interpolation blocks; reserved for contouring.
G02, G03	Used with circular interpolation.
G04	Sets a calculated time delay during which there is no machine motion (dwell).
G05, G07	Unassigned by the EIA. May be used at the discretion of the machine tool or system builder. Could also be standardized at a future date.
G06	Used with parabolic interpolation.
G08	Acceleration code. Causes the machine, assuming it is capable, to accelerate at a smooth exponential rate.
G09	Deceleration code. Causes the machine, assuming it is capable, to decelerate at a smooth exponential rate.
G10–G12	Normally unassigned for CNC systems. Used with some hard-wired systems to express blocks of abnormal dimensions.
G13–G16	Direct the control system to operate on a particular set of axes.
G17–G19	Identify or select a coordinate plane for such functions as circular interpolation or cutter compensation.
G20–G32	Unassigned according to EIA standards. May be assigned by the control system or machine tool builder.
G33–G35	Selected for machines equipped with thread-cutting capabilities (generally referring to lathes). G33 is used when a constant lead is sought, G34 is used when a constantly increasing lead is required, and G35 is used to designate a constantly decreasing lead.
G36–G39	Unassigned.
G40	Terminates any cutter compensation.
G41	Activates cutter compensation in which the cutter is on the left side of the work surface (relative to the direction of the cutter motion).
G42	Activates cutter compensation in which the cutter is on the right side of the work surface.

G43, G44   Used with cutter offset to adjust for the difference between the actual and programmed cutter radii or diameters. G43 refers to an inside corner, and G44 refers to an outside corner.

G45–G49   Unassigned.

G50–G59   Reserved for adaptive control.

G60–G69   Unassigned.

G70   Selects inch programming.

G71   Selects metric programming.

G72   Selects three-dimensional CW circular interpolation.

G73   Selects three-dimensional CCW circular interpolation.

G74   Cancels multiquadrant circular interpolation.

G75   Activates multiquadrant circular interpolation.

G76–G79   Unassigned.

G80   Cancel cycle.

G81   Activates drill, or spotdrill, cycle.

G82   Activates drill with a dwell.

G83   Activates intermittent, or deep-hole, drilling.

G84   Activates tapping cycle.

G85–G89   Activate boring cycles.

G90   Selects absolute input. Input data is to be in absolute dimensional form.

G91   Selects incremental input. Input data is to be in incremental form.

G92   Preloads registers to desired values (for example, preloads axis position registers).

G93   Sets inverse time feed rate.

G94   Sets inches (or millimeters) per minute feed rate.

G95   Sets inches (or millimeters) per revolution feed rate.

G97   Sets spindle speed in revolutions per minute.

G98, G99   Unassigned.

Miscellaneous Functions

M word	Explanation
M00	Program stop. Operator must cycle start in order to continue with the remainder of the program.
M01	Optional stop. Acted upon only when the operator has previously signaled for this command by pushing a button. When the control system senses the M01 code, machine will automatically stop.
M02	End of program. Stops the machine after completion of all commands in the block. May include rewinding of tape.
M03	Starts spindle rotation in a clockwise direction.
M04	Starts spindle rotation in a counterclockwise direction.
M05	Spindle stop.

M06	Executes the change of a tool (or tools) manually or automatically.
M07	Turns coolant on (flood).
M08	Turns coolant on (mist).
M09	Turns coolant off.
M10	Activates automatic clamping of the machine slides, workpiece, fixture, spindle, etc.
M11	Deactivates automatic clamping.
M12	Inhibiting code used to synchronize multiple sets of axes, such as a four-axis lathe that has two independently operated heads or slides.
M13	Combines simultaneous clockwise spindle motion and coolant on.
M14	Combines simultaneous counterclockwise spindle motion and coolant on.
M15	Sets rapid traverse or feed motion in the + direction.
M16	Sets rapid traverse or feed motion in the − direction.
M17, M18	Unassigned.
M19	Oriented spindle stop. Stops spindle at a predetermined angular position.
M20–M29	Unassigned.
M30	End of data. Used to reset control and/or machine.
M31	Interlock bypass. Temporarily circumvents a normally provided interlock.
M32–M39	Unassigned.
M40–M46	Signals gear changes if required at the machine; otherwise, unassigned.
M47	Continues program execution from the start of the program, unless inhibited by an interlock signal.
M48	Cancels M49.
M49	Deactivates a manual spindle or feed override and returns to the programmed value.
M50–M57	Unassigned.
M58	Cancels M59.
M59	Holds the RPM constant at its value.
M60–M99	Unassigned.

Other Address Characters

Address character	Explanation
A	Angular dimension about the X-axis.
B	Angular dimension about the Y-axis.
C	Angular dimension about the Z-axis.
D	Can be used for an angular dimension around a special axis, for a third feed function, or for tool offset.
E	Used for angular dimension around a special axis or for a second feed function.
H	Fixture offset.
I, J, K	Centerpoint coordinates for circular interpolation.

L	Not used.
O	Used on some N/C controls in place of the customary sequence number word address N.
P	Third rapid traverse code—tertiary motion dimension parallel to the X-axis.
Q	Second rapid traverse code—tertiary motion dimension parallel to the Y-axis.
R	First rapid traverse code—tertiary motion dimension parallel to the Z-axis (or to the radius) for constant surface speed calculation.
U	Secondary motion dimension parallel to the X-axis.
V	Secondary motion dimension parallel to the Y-axis.
W	Secondary motion dimension parallel to the Z-axis.

APPENDIX B _____

GENERAL SAFETY RULES FOR N/C MACHINES

1. Wear safety glasses at all times.
2. Wear safety shoes.
3. Do not wear neckties, long sleeves, wristwatches, rings, gloves, etc., when operating machine.
4. Keep long hair covered when operating machine.
5. Make sure the area around the machine is well lighted, dry, and as free from obstructions as possible. Keep the area in good order.
6. Never perform grinding operations near an N/C machine. Abrasive dust will cause undue wear, inaccuracies, and possible failure of affected parts.
7. Do not use compressed air to blow chips from the part, machine surfaces, cabinets, controls, or floor around the machine.
8. When handling or lifting parts or tooling, follow company policy on correct procedures.
9. Work platforms around machines should be sturdy and must have anti-slip surfaces.
10. Wrenches, tools, and other parts should be kept off the machine and all its moving units. Do not use machine elements as a workbench.
11. Keep hands out of path of moving units during machining operations.
12. Never place hands near a revolving spindle.
13. Perform all setup work with spindle stopped.
14. Load and unload workpieces with spindle stopped.
15. Securely clamp all work and fixtures before starting machine.
16. When handling tools or changing tools by hand, use a glove or cloth. Avoid contact with cutting edges. Do not operate machine with gloves.
17. Use caution when changing tools, and avoid interference with fixture or workpiece.
18. Use only properly sharpened tools.
19. Clean the setup daily.
20. Avoid bumping any N/C machine or controls.

21. Never operate an N/C machine or any other automated equipment without consulting the specific operator's manual for that particular machine and control type.

22. Never attempt to program an N/C machine or any other automated equipment without consulting the specific programmer's manual for that particular machine and control type.

23. Electrical compartment doors should be opened *only* for electrical and/or maintenance work. They should be opened only by experienced electricians and/or qualified service personnel.

24. Safety guards, covers, and other devices have been provided for protection. Do not operate machine with these devices disconnected, removed, or out of place. Operate machine only when these items are in proper operating condition and position.

25. Tools are made for right-hand or left-hand operation. Be sure that spindle direction is correct.

26. Do not remove chips from workpiece area with fingers or while spindle is running. Use a brush to remove chips *after* the spindle has stopped. Clear chips often.

27. Pay attention to all warning lights, beacons, and audible signals.

28. Never attempt to alter automated cell or system hardware or software without the proper authorization and approval.

APPENDIX C _____

USEFUL FORMULAS AND TABLES

Determining RPM

$$RPM = \frac{3.82 \times Cutspeed}{Diameter}$$

Switching between feed values in inches per minute (IPM) and inches per revolution (IPR)

$$Feed\ (IPM) = (RPM) \times Feed\ (IPR)$$

$$Feed\ (IPR) = \frac{Feed\ (IPM)}{RPM}$$

Useful calculations

- To find the circumference of a circle, multiply the diameter by 3.1416.
- To find the diameter of a circle, multiply the circumference by 0.31831.
- To find the area of a circle, multiply the square of the diameter by 0.7854.
- To obtain the circumference, multiply the radius of a circle by 6.283185.
- To obtain the area of a circle, multiply the square of the circumference of a circle by 0.07958.
- To find the area of a circle, multiply half the circumference of a circle by half its diameter.
- To obtain the radius of a circle, multiply the circumference of a circle by 0.159155.
- To find the radius of a circle, multiply the square root of the area of a circle by 0.56419.
- To find the diameter of a circle, multiply the square root of the area of a circle by 1.12838.
- To find the area of the surface of a ball (sphere), multiply the square of the diameter by 3.1416.
- To find the volume of a ball (sphere), multiply the cube of the diameter by 0.5236.

Trigonometry

$$c^2 = a^2 + b^2$$

$$c = \sqrt{a^2 + b^2} \quad a = \sqrt{c^2 - b^2} \quad b = \sqrt{c^2 - a^2}$$

where c = hypotenuse length of a right triangle

a = length of one of the short sides of a right triangle

b = length of the other short side

$$\text{Sine} = \frac{\text{Side opposite}}{\text{Hypotenuse}} \qquad \text{Cosecant} = \frac{\text{Hypotenuse}}{\text{Side opposite}}$$

$$\text{Cosine} = \frac{\text{Side adjacent}}{\text{Hypotenuse}} \qquad \text{Secant} = \frac{\text{Hypotenuse}}{\text{Side adjacent}}$$

$$\text{Tangent} = \frac{\text{Side opposite}}{\text{Side adjacent}} \qquad \text{Cotangent} = \frac{\text{Side adjacent}}{\text{Side opposite}}$$

CUTTING SPEEDS
(Feet Per Minute)

MATERIAL	DRILL HSS	DRILL CARBIDE	REAM HSS	REAM CARBIDE	TAP HSS	COBORE HSS	COBORE CARBIDE	BORE HSS	BORE CARBIDE	MILLING HIGH SPEED Rgh.	MILLING HIGH SPEED Fin.	MILLING CARBIDE Rgh.	MILLING CARBIDE Fin.
Aluminum	200	350	175	300	90	180	300	300	600	240	300	500	1000
Brass — Soft	145	350	120	250	100	150	300	150	450	150	200	400	600
Brass — Hard	125	225	100	200	75	110	200	120	350	135	180	350	500
Bronze — Common	140	250	125	200	90	130	200	150	400	145	190	360	550
Bronze — High Tensile	60	200	50	175	40	55	180	85	300	70	90	200	280
Cast Iron — Soft 170 BHN	90	180	60	200	40	85	160	80	280	90	110	250	350
Cast Iron — Medium 220 BHN	60	140	45	125	30	55	130	55	255	70	90	200	300
Cast Iron — Hard 300 BHN	40	120	30	60	20	35	100	45	215	50	60	175	250
Cast Iron — Malleable	85	140	45	100	40	75	180	90	250	100	120	260	370
Cast Steel	60	120	50	100	40	60	180	70	200	50	80	225	380
Copper	75	250	50	125	40	70	200	95	350	90	150	220	400
Magnesium	250	500	180	450	150	200	450	400	1000	300	400	600	1000
Monel	50	100	35	90	20	45	90	50	110	60	80	180	240
Steel — Mild .2 to .3 Carbon	95	—	50	250	40	85	170	80	280	90	130	300	450
Steel — Medium .4 to .5 Carbon	75	—	45	200	35	60	120	80	220	70	85	210	400
Steel — Tool up to 1.2 Carbon	40	80	30	70	20	40	80	45	190	50	80	175	350
Steel — Forging	45	90	35	80	25	40	80	50	200	60	80	200	300
Steel — Alloy 300 BHN	60	120	40	115	35	60	120	70	250	60	80	250	350
Steel — Alloy 400 BHN	45	90	30	65	25	40	80	40	165	30	40	160	250
Steel — High Tensile to 40 R_c	35	70	30	60	20	30	60	40	150	40	50	120	150
Steel — High Tensile to 45 R_c	30	60	20	50	15	20	40	30	100	35	45	110	140
Steel — Stainless — Free Machining	55	110	35	100	25	50	100	50	150	40	60	200	400
Steel — Stainless — Work Hardening	30	60	20	50	15	30	60	40	90	30	50	180	300
Titanium — Commercially Pure	55	110	45	100	30	50	100	60	120	60	75	200	280
Zinc Die Casting	150	300	125	225	80	150	250	180	350	200	300	250	450

CONVERSION CHART
CUTTING SPEEDS TO RPM

CUTTING SPEEDS (left column) vs. TOOL DIAMETERS (across top)

CUTTING SPEED	1/8	3/16	1/4	5/16	3/8	7/16	1/2	9/16	5/8	11/16	3/4	13/16	7/8	15/16	1	1-1/8	1-1/4	1-3/8	1-1/2	1-5/8	1-3/4	2	2-1/4	2-1/2	2-3/4	3	3-1/2	4	4-1/2	5	5-1/2	6	6-1/2	7	7-1/2	8
10	306	204	153	122	102	87	76	68	61	56	51	47	44	41	38	34	31	28	25	24	22	19	17	15												
20	611	407	306	244	204	175	153	136	122	111	102	94	87	82	76	68	61	56	51	47	44	38	34	31	28	25	22	19	17	15						
30	917	611	458	368	306	262	(229)	204	183	167	153	141	131	122	115	102	92	83	76	71	65	57	51	46	42	38	33	29	25	23	21	19	18	16	15	
40	1222	815	611	489	408	349	306	272	245	222	204	188	175	163	153	136	122	111	102	94	87	76	68	61	56	51	44	38	34	31	28	25	24	22	20	19
50	1528	1020	764	611	509	437	382	339	306	278	255	235	218	204	191	170	153	140	127	118	109	95	85	76	69	64	55	48	42	38	35	32	29	27	25	24
60	1834	1222	917	733	611	524	458	407	367	333	306	282	262	245	229	204	183	167	153	141	131	(115)	102	92	83	76	65	57	51	46	42	38	35	33	31	29
70	2140	1426	1070	856	713	611	535	475	428	389	357	329	306	285	267	238	214	194	178	165	153	134	119	107	97	89	76	67	59	53	49	45	41	38	36	33
80	2445	1630	1222	978	813	700	611	543	489	444	408	376	350	326	306	272	244	222	204	188	175	153	136	122	111	102	87	76	68	61	56	51	47	44	41	38
90	2750	1833	1375	1100	917	786	688	611	550	500	458	423	393	367	344	306	275	250	229	212	196	172	153	138	125	115	98	86	76	69	63	57	53	49	46	43
100		2037	1528	1222	1020	873	764	679	611	556	509	470	436	408	382	340	306	278	255	235	218	191	170	153	139	127	109	96	85	76	70	64	59	55	51	48
120		2445	1834	1467	1222	1048	917	815	733	667	611	564	524	489	458	407	367	333	306	282	262	229	204	183	167	153	131	115	102	92	83	76	71	65	61	57
140		2852	2140	1711	1426	1222	1070	950	856	778	713	658	611	571	535	475	428	390	356	329	306	267	238	214	194	178	153	134	119	107	97	89	82	76	71	67
150			2292	1834	1528	1310	1146	1018	917	833	764	705	655	611	573	509	458	417	382	353	327	286	255	229	208	191	164	143	127	115	104	95	88	82	76	72
160			2445	1956	1630	1397	1222	1086	978	889	815	752	698	652	611	543	490	444	407	376	350	306	272	244	222	204	175	153	136	122	111	102	94	87	81	76
180			2750	2200	1834	1572	1375	1222	1100	1000	917	846	786	734	688	611	550	500	458	423	393	344	306	275	250	229	196	172	153	138	125	115	106	98	92	86
200				2445	2037	1747	1528	1358	1222	1111	1020	940	874	815	764	680	611	556	510	470	437	382	340	306	278	255	218	(191)	170	153	139	127	117	109	102	96
220				2690	2240	1920	1681	1494	1345	1222	1121	1034	960	897	840	747	672	611	560	517	480	420	374	336	306	280	240	210	187	168	153	140	129	120	112	105
240				2934	2445	2096	1833	1630	1467	1333	1222	1128	1048	978	917	815	733	667	611	564	524	458	407	367	333	306	262	229	204	183	167	153	141	131	122	115
250					2547	2183	1910	1697	1528	1389	1274	1175	1091	1020	955	850	764	694	637	588	546	477	424	382	347	318	273	240	212	191	174	159	147	136	127	119
260					2650	2270	1986	1765	1590	1444	1325	1221	1135	1060	993	883	795	722	662	611	568	497	441	397	361	331	284	250	221	200	181	166	153	142	132	124
280					2850	2445	2140	1900	1712	1556	1426	1316	1222	1140	1070	950	856	778	713	658	611	535	475	428	390	357	306	267	238	214	194	178	165	153	143	134
300						2620	2292	2037	1834	1667	1528	1410	1310	1222	1146	1020	917	833	764	705	655	573	510	458	417	382	327	287	255	229	208	191	176	164	153	143
320						2795	2445	2172	1956	1778	1630	1504	1397	1304	1222	1086	978	889	815	752	698	611	543	490	444	407	350	306	272	244	222	204	188	175	163	153
340						2970	2597	2308	2078	1889	1732	1600	1484	1385	1300	1155	1040	944	866	800	742	650	577	520	472	433	372	325	290	260	236	216	200	186	173	162
350							2674	2375	2140	1944	1783	1646	1528	1426	1337	1190	1070	972	891	823	764	668	594	535	486	446	381	334	297	267	243	223	206	191	178	167
360							2750	2444	2200	2000	1834	1693	1570	1467	1375	1222	1100	1000	917	846	786	688	611	550	500	458	393	344	306	275	250	230	212	196	183	172
380							2900	2580	2323	2111	1936	1787	1660	1550	1450	1290	1160	1055	968	893	830	725	645	580	528	484	415	363	323	290	264	242	223	207	194	181
400								2715	2445	2222	2038	1881	1746	1630	1530	1360	1222	1111	1020	940	873	764	680	611	556	510	437	382	340	305	278	255	235	218	204	191
450									2750	2500	2292	2116	1964	1834	1720	1530	1375	1250	1146	1060	982	860	764	688	625	573	491	430	382	344	313	286	264	246	230	215
500										2778	2550	2350	2180	2040	1910	1700	1530	1390	1273	1175	1090	955	850	764	694	637	546	478	425	382	348	318	294	273	255	239
550											2800	2586	2400	2240	2100	1870	1680	1530	1400	1293	1200	1050	934	840	764	700	600	525	467	420	382	350	325	300	280	263
600												2821	2620	2445	2290	2040	1834	1667	1528	1410	1310	1145	1020	917	833	764	655	573	510	458	417	382	353	327	306	287

TAPPING FEED RATES

THREADS PER INCH-TO-LEAD

TPI	LEAD	TPI	LEAD	TPI	LEAD
3	.3333	11-1/2	.0870	32	.0313
3-1/2	.2587	12	.0833	36	.0278
4	.2500	13	.0769	40	.0250
5	.2000	14	.0714	44	.0227
6	.1667	16	.0625	48	.0208
7	.1430	18	.0556	56	.0179
8	.1250	20	.0500	64	.0156
9	.1111	24	.0417	72	.0139
10	.1000	27	.0370	80	.0125
11	.0909	28	.0357		

Program feed rate = Lead of tap (inches) × RPM

CONVERSION CHART
(Based on 25.4 mm = 1″)
Inches into Millimeters

Inches	M/M	Inches		M/M	Inches	M/M	Inches	M/M	Inches	M/M
1/64	.0156	0.3969	49/64	.7656	19.4469	34	863.600	82	2082.80	130	3302.00
1/32	.0313	0.7937	25/32	.7813	19.8437	35	889.000	83	2108.20	131	3327.40
3/64	.0469	1.1906	51/64	.7969	20.2406	36	914.400	84	2133.60	132	3352.80
1/16	.0625	1.5875	13/16	.8125	20.6375	37	939.800	85	2159.00	133	3378.20
5/64	.0781	1.9844	53/64	.8281	21.0344	38	965.200	86	2184.40	134	3403.60
3/32	.0938	2.3812	27/32	.8438	21.4312	39	990.600	87	2209.80	135	3429.00
7/64	.1094	2.7781	55/64	.8594	21.8281	40	1016.00	88	2235.20	136	3454.40
1/8	.1250	3.1750	7/8	.8750	22.2250	41	1041.40	89	2260.60	137	3479.80
9/64	.1406	3.5719	57/64	.8906	22.6219	42	1066.80	90	2286.00	138	3505.20
5/32	.1563	3.9687	29/32	.9063	23.0187	43	1092.20	91	2311.40	139	3530.60
11/64	.1719	4.3656	59/64	.9219	23.4156	44	1117.60	92	2336.80	140	3556.00
3/16	.1875	4.7625	15/16	.9375	23.8125	45	1143.00	93	2362.20	141	3581.40
13/64	.2031	5.1594	61/64	.9531	24.2094	46	1168.40	94	2387.60	142	3606.80
7/32	.2188	5.5562	31/32	.9688	24.6062	47	1193.80	95	2413.00	143	3632.20
15/64	.2344	5.9531	63/64	.9844	25.0031	48	1219.20	96	2438.40	144	3657.60
1/4	.2500	6.3500	1		25.4000	49	1244.60	97	2463.80	145	3683.00
17/64	.2656	6.7469	2		50.800	50	1270.00	98	2489.20	146	3708.40
9/32	.2813	7.1437	3		76.200	51	1295.40	99	2514.60	147	3733.80
19/64	.2969	7.5406	4		101.600	52	1320.80	100	2540.00	148	3759.20
5/16	.3125	7.9375	5		127.000	53	1346.20	101	2565.40	149	3784.60
21/64	.3281	8.3344	6		152.400	54	1371.60	102	2590.80	150	3810.00
11/32	.3438	8.7312	7		177.800	55	1397.00	103	2616.20	151	3835.40
23/64	.3594	9.1281	8		203.200	56	1422.00	104	2641.60	152	3860.80
3/8	.3750	9.5250	9		228.600	57	1447.80	105	2667.00	153	3886.20
25/64	.3906	9.9219	10		254.000	58	1473.20	106	2692.40	154	3911.60
13/32	.4063	10.3187	11		279.400	59	1498.60	107	2717.80	155	3937.00
27/64	.4219	10.7156	12		304.800	60	1524.00	108	2743.20	156	3962.40
7/16	.4375	11.1125	13		330.200	61	1549.40	109	2768.60	157	3987.80
29/64	.4531	11.5094	14		355.600	62	1574.80	110	2794.00	158	4013.20
15/32	.4688	11.9062	15		381.000	63	1600.20	111	2819.40	159	4038.60
31/64	.4844	12.3031	16		406.400	64	1625.60	112	2844.80	160	4064.00
1/2	.5000	12.7000	17		431.800	65	1651.00	113	2870.20	161	4089.40
33/64	.5156	13.0969	18		457.200	66	1676.40	114	2895.60	162	4114.80
17/32	.5313	13.4937	19		482.600	67	1701.80	115	2921.00	163	4140.20
35/64	.5469	13.8906	20		508.000	68	1727.20	116	2946.40	164	4165.60
9/16	.5625	14.2875	21		533.400	69	1752.60	117	2971.80	165	4191.00
37/64	.5781	14.6844	22		558.800	70	1778.00	118	2997.20	166	4216.40
19/32	.5938	15.0812	23		584.200	71	1803.40	119	3022.60	167	4241.80
39/64	.6094	15.4781	24		609.600	72	1828.80	120	3048.00	168	4267.20
5/8	.6250	15.8750	25		635.000	73	1854.20	121	3073.40	169	4292.60
41/64	.6406	16.2719	26		660.400	74	1879.60	122	3098.80	170	4318.00
21/32	.6563	16.6687	27		685.800	75	1905.00	123	3124.20	171	4343.40
43/64	.6719	17.0656	28		711.200	76	1930.40	124	3149.60	172	4368.80
11/16	.6875	17.4625	29		736.600	77	1955.80	125	3175.00	173	4394.20
45/64	.7031	17.8594	30		762.000	78	1981.20	126	3200.40	174	4419.60
23/32	.7188	18.2562	31		787.400	79	2006.60	127	3225.80	175	4445.00
47/64	.7344	18.6531	32		812.800	80	2032.00	128	3251.20		
3/4	.7500	19.0500	33		838.200	81	2057.40	129	3276.60		

0.001″ = .0254 mm   0.001 mm = 0.0004″

GLOSSARY

A-AXIS. (or α-axis) The axis of circular motion of a machine tool member or slide about the X-axis. Values along the A-axis are degrees of rotation about the X-axis.

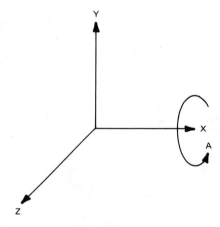

ABSOLUTE ACCURACY. Accuracy as measured from a specified reference.

ABSOLUTE READOUT. A display of the true slide position as derived from the position commands within the control system.

ABSOLUTE SYSTEM. A numerical control system in which all positional dimensions, both input and feedback, are given with respect to a common datum point. The alternative is the incremental system.

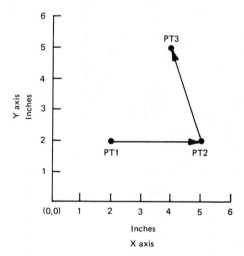

Coordinate Positions

Point	X value	Y value
PT1	2	2
PT2	5	2
PT3	4	5

In an absolute system, all points are relative to (0,0), and the absolute coordinates for each of the required points are programmed with respect to (0,0).

306

ACCANDEC. (acceleration and deceleration) Acceleration and deceleration in feed rate. It provides smooth starts and stops when operating in N/C and when changing from one feed rate value to another.

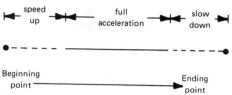

Most modern numerical control systems have automatic acceleration and deceleration.

ACCURACY. 1. A measure of the difference between the actual position of the machine slide and the position demanded. 2. Conformity of an indicated value to a true value (that is, an actual or an accepted standard value). The accuracy of a control system is expressed as the deviation (the difference between the ultimately controlled variable and its ideal value), usually in the steady state or at sampled instants.

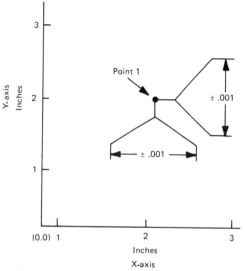

The position of point 1 in this example is X = 2 and Y = 2. If the machine accuracy is specified as ± .001, the X-axis movement could be between X = 1.999 and X = 2.001. The Y-axis movement could be between Y = 1.999 and Y = 2.001.

C1 = circle/center, PT1, radius, 2.5

Similar to the APT language except it does not possess the advanced contouring capabilities of APT.

AD-APT. An Air Force adaptation of APT program language, with limited vocabulary. It can be used for N/C programming on some small- to medium-size U.S. computers.

ADAPTIVE CONTROL. A technique that automatically adjusts feeds and/or speeds to an optimum by sensing cutting conditions and acting upon them.

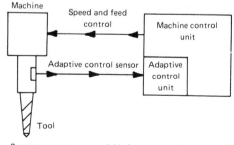

Sensors may measure variable factors, e.g. vibration, heat, torque, and deflection. Cutting speeds and feeds may be increased or decreased depending on conditions sensed.

ADDRESS. 1. A symbol indicating the significance of the information immediately following. 2. A means of identifying information or a location in a control system. 3. A number that identifies one location in memory.

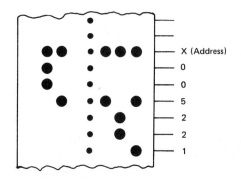

ALGORITHM. 1. A computational procedure for solving a problem. When properly applied, an algorithm always produces a solution to a problem.

ALPHANUMERIC CODING. A system in which the characters are letters A through Z and numerals 0 through 9.

APT and AD-APT statements use alphanumeric coding, e.g. GOFWD, CT12/PAST, 2, INTOF, L13

ANALOG. 1. Pertaining to a system that uses electrical voltage magnitudes or ratios to represent physical axis positions. 2. Pertaining to information that can have continuously variable values.

ANALYST. A person skilled in the definition and development of techniques to solve problems.

APT. (Automatic Programmed Tool) A universal computer-assisted program system for multiaxis contouring programming. APT III provides for five axes of machine tool motion.

Typical APT geometry definition statement:
C1 = CIRCLE/XLARGE, L12, XLARGE, L13, RADIUS, 3.5

Typical APT tool motion statement:
TLRGT, GORGT/AL3, PAST, AL12

ARC CLOCKWISE. An arc, generated by the coordinated motion of two axes, in which curvature of the tool path with respect to the workpiece is clockwise when the plane of motion is viewed from the positive direction of the perpendicular axis.

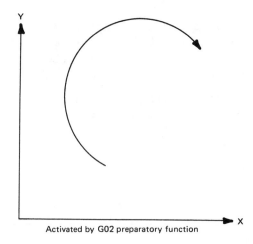

Activated by G02 preparatory function

ARC COUNTERCLOCKWISE. An arc, generated by the coordinated motion of two axes, in which curvature of the tool path with respect to the workpiece is counterclockwise when the plane of motion is viewed from the positive direction of the perpendicular axis.

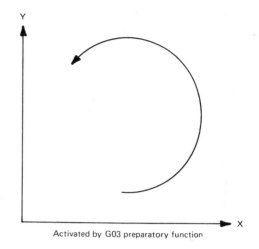

Activated by G03 preparatory function

ARCHITECTURE. The way in which the parts of a system fit and communicate with one another.

ASCII. (American Standard Code for Information Interchange) A data transmission code that has been established as an American standard by the American Standards Association. It is a code in which seven bits are used to represent each character. Formerly USASCII.

AUTOMATION. 1. The implementation of processes by automatic means. 2. The investigation, design, development, and application of methods to render processes automatic, self-moving, or self-controlling.

AUTOSPOT. (Automatic System for Positioning of Tools) An older computer-assigned program for N/C positioning and straight-cut systems, developed in the United States by the IBM Space Guidance Center. At one time, it was maintained and taught by IBM.

AUXILIARY FUNCTION. A programmable function of a machine other than the control of the coordinate movements or cutter.

- Transferring a tool to the select tool position.
- Turning coolant ON or OFF.
- Starting or stopping the spindle.
- Initiating pallet shuttle or movement.

AXIS. A principal direction along which the relative movements of the tool or workpiece occur. There are usually three linear axes, mutually at right angles, designated as X, Y, and Z.

AXIS INHIBIT. A feature of an N/C unit that enables the operator to withhold command information from a machine tool slide.

AXIS INTERCHANGE. The entering of information concerning one axis into the storage location of another axis.

AXIS INVERSION. The reversal of plus and minus values along an axis. This allows the machining of a left-handed part from right-handed programming, or vice versa.

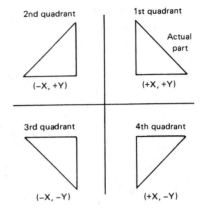

BACKLASH. A relative movement between interacting mechanical parts as a result of looseness.

BATCH PROCESSING. Technique by which items to be processed must be coded and collected into groups prior to processing.

B-AXIS. (or β-axis) The axis of circular motion of a machine tool member or slide about the Y-axis.

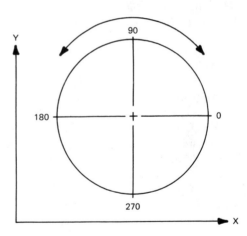

BCD. (binary-coded decimal) A system of number representation in which each decimal digit is represented by a group of binary digits that corresponds to a character.

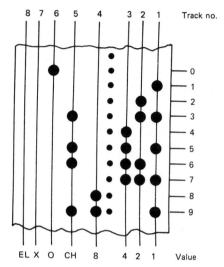

Numbers and letters are expressed by punched holes across the tape for the code or value desired.

BINARY CODE. A code based on binary numbers, which are expressed as either 1 or 0, true or false, on or off.

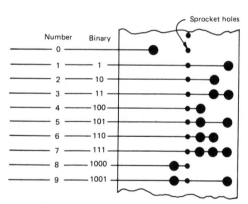

Most computers operate on some form of binary system where a number or letter can be expressed as ON (a hole) or OFF (no hole).

BIT. (binary digit) 1. A binary digit having only two possible states. 2. A single character of a language that uses exactly two distinct kinds of characters. 3. A magnetized spot on any storage device.

BLOCK. A word or group of words considered as a unit. A block is separated from other units by an end-of-block character. On punched tape, a block of data provides sufficient information for an operation.

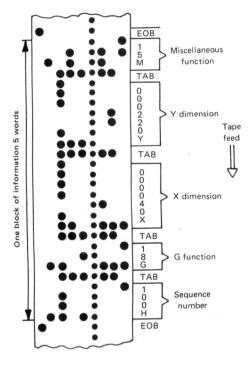

BLOCK DELETE. A function that permits selected blocks of tape to be ignored by the control system, at the operator's discretion, with permission of the programmer.

This feature allows certain blocks of information to be skipped by programming a slash (/) code in front of the block to be skipped. One lot of parts with holes 1, 2, and 3 are required. On another lot, only holes 1 and 3 are required. The same tape could be used for both lots by activating the block delete switch on the second lot and eliminating hole 2. The (/) code would be in front of the block of information for hole 2.

BUFFER STORAGE. A place in which information in a control system or computer is stored for planned use. Information from the buffer storage section of a control system can be transferred almost instantly to active storage (the portion of the control system commanding the operation at that particular time). Buffer storage allows a control system to act immediately on stored information, rather than wait for the information to be read into the machine from the tape reader.

BUG. 1. A mistake or malfunction. 2. An integrated circuit (slang).

BYTE. A sequence of adjacent binary digits usually operated on as a unit and shorter than a computer word.

Eight bits equal one byte. A computer word usually consists of either sixteen or thirty-two bits (two or four bytes).

CAD. (Computer-aided design) The use of computers to assist in phases of design.

CAM. (Computer-aided manufacturing) The use of computers to assist in phases of manufacturing.

CAM-I. (Computer-Aided Manufacturing International) The outgrowth and replacement organization of the APT Long Range Program.

CANCEL. A command that will discontinue any canned cycles or sequence commands.

CANNED CYCLE. A preset sequence of events initiated by a single command. For example, code G84 will perform a tapping cycle by N/C.

CARD-TO-TAPE CONVERTER. A device that converts information directly from punched cards to punched or magnetic tape.

CARTESIAN COORDINATES. A three-dimensional system whereby the position of a point can be defined with reference to a set of axes at right angles to each other.

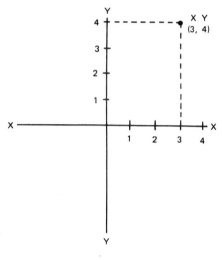

C-AXIS. Normally the axis of circular motion of a machine tool member or slide about the Z-axis. Values along the C-axis are degrees of rotation about the Z-axis.

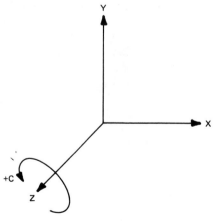

CHAD. Pieces of material removed in card or tape operations.

CHANNELS. Paths parallel to the edge of the tape, along which information may be stored by the presence or absence of holes or magnetized areas. This term is also known as *level* or *track*. The EIA standard 1-in.-wide tape has eight channels.

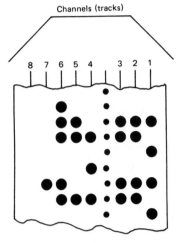

CHARACTERS. A general term for all symbols, such as alphabetic letters, numerals, and punctuation marks. It is also the coded representation of such symbols.

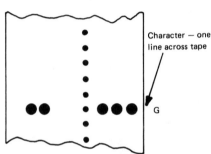

CHIP. A single piece of silicon cut from a slice by scribing and breaking. It can contain one or more circuits but is packaged as a unit.

CIRCULAR INTERPOLATION. 1. The process of generating up to 360 degrees of arc using only one block of information as defined by EIA. 2. A mode of contouring control that uses the information contained in a single block to produce an arc of a circle.

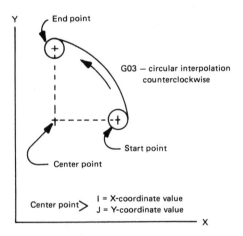

CL DATA. Processor output that contains information regarding cutter location.

CLOSED-LOOP SYSTEM. A system in which the output, or some result of the output, is measured and fed back for comparison with the input. In an N/C system, the output is the position of the table or head; the input is the tape information, which ordinarily differs from the output. This difference is measured and results in a machine movement to reduce and eliminate the variance.

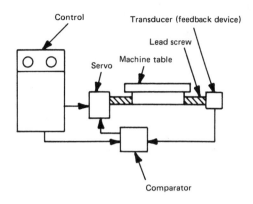

CNC. Computer numerical control

CODE. A system describing the formation of characters on a tape; it is used to represent information and is written in a language that can be understood and handled by the control system.

COMMAND. A signal or series of signals that initiates one step in the execution of a program.

COMMAND READOUT. A display of the slide position as commanded from the control system.

COMPILE. To generate a machine language program from a computer program written in a high-level source code.

COMPILER. Computer program used to translate high-level source code into machine language programs suitable for execution on a particular computing system.

CONSOLE. That part of a computer used for communication between the computer operator or maintenance engineer and the computer.

CONTINUOUS-PATH OPERATION. An operation in which rate and direction of relative movement of machine members is under continuous numerical control. There is no pause for data reading. (See *contouring control system*)

CONTOURING CONTROL SYSTEM. An N/C system for controlling a machine (for example, during milling or drafting) in a path that results from the coordinated, simultaneous motion of two or more axes.

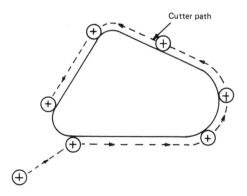

COORDINATE DIMENSIONING. A system of dimensioning based on a common starting point.

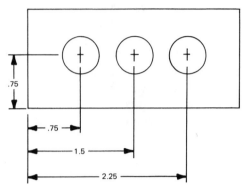

(Also known as absolute dimensioning)

CPU. Central processing unit of a computer. The memory or logic of a computer that includes overall circuits, processing, and execution of instructions.

CRT. (cathode ray tube) A device that represents data (alphanumeric or graphic) by means of a controlled electron beam directed against a fluorescent coating in the tube.

CUTTER DIAMETER COMPENSATION. A system in which the programmed path may be altered to allow for the difference between actual and programmed cutter diameters.

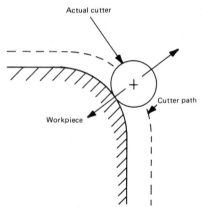

CUTTER OFFSET. The distance from the part surface to the axial center of a cutter (that is, the radius of the cutting tool).

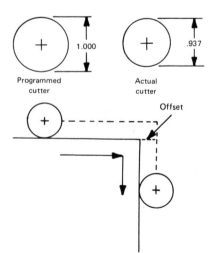

CUTTER PATH. The path defined by the center of a cutter.

CYCLE. 1. A sequence of operations that is regularly repeated. 2. The time it takes for one such sequence to occur.

DATA. A representation of information in the form of words, symbols, numbers, letters, characters, digits, etc.

DATABASE. A comprehensive data file containing information in a format applicable to a user's needs and available when needed.

DEBUG. 1. To detect, locate, and remove mistakes from a program. 2. Troubleshoot.

DECIMAL CODE. A code in which each allowable position has one of ten possible states. (The conventional decimal number system is a decimal code.)

DELETE CHARACTER. A character used primarily to obliterate any erroneous or unwanted characters on punched tape. The delete character consists of perforations in all punching positions.

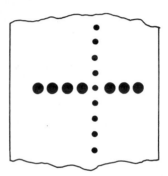

DIAGNOSTIC TEST. The running of a machine program or routine to discover a failure or potential failure of a machine element and to determine its location.

DIGIT. A character in any numbering system.

DIGITAL. 1. Referring to discrete states of a signal (on or off) (a combination of these makes up a specific value). 2. Relating to data in the form of digits.

DISPLAY. A visual representation of data.

DOCUMENTATION. Manuals and other printed materials (tables, magnetic tape, listing, diagrams) that provide information for the use and maintenance of a manufactured product, both hardware and software.

DOWNTIME. Time during which equipment is inoperable because of faults.

DNC. Direct or distributed numerical control.

DWELL TIME. A timed delay of programmed or established duration, not cyclic or sequential. It is not an interlock or hold time.

EDIT. To modify the form of data.

EIA STANDARD CODE. A standard code for positioning, straight-cut, and contouring control systems proposed by the U.S. EIA in their Standard RS-244. Eight-track paper (1-in. wide) has been accepted by the American Standards Association as an American standard for numerical control.

END-OF-BLOCK CHARACTER. 1. A character indicating the end of a block of tape information, used to stop the tape reader after a block has been read. 2. The typewriter function of the carriage return during preparation of machine control tapes.

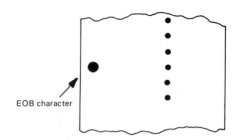

EOB character

END OF PROGRAM. A miscellaneous function (M02) that indicates the completion of a workpiece. Stops spindle, coolant, and feed after completion of all commands in the block. Used to reset control and/or machine.

END OF TAPE. A miscellaneous function (M30) that stops spindle, coolant, and feed after completion of all commands in the block. Used to reset control and/or machine.

END POINTS. The extremities of a span.

ERROR SIGNAL. Indication of a difference between the output and input signals in a servo system.

EXECUTIVE PROGRAM. A series of programming instructions enabling a dedicated minicomputer to produce a specific output control. For example, it is the executive program in a CNC unit that enables the control to think like a lathe or machining center.

FEED. The programmed or manually established rate of movement of the cutting tool into the workpiece for the required machining operation.

FEEDBACK. The transmission of a signal from a late to an earlier stage in a system. In a closed-loop N/C system, a signal of the machine slide position is fed back and compared with the input signal, which specifies the demanded position. These two signals are compared, and an error signal is generated if a difference exists.

FEED FUNCTION. The relative motion between the tool or instrument and the work due to motion of the programmed axis.

FEED RATE (CODE WORD). A multiple-character code containing the letter F followed by digits. It determines the machine slide rate of feed.

FEED RATE DIVIDER. A feature of some machine control units that allows the programmed feed rate to be divided by a selected amount as provided for in the machine control unit.

FEED RATE MULTIPLIER. A feature of some machine control units that allows the programmed feed rate to be multiplied by a selected amount as provided for in the machine control unit.

FEED RATE OVERRIDE. A variable manual control function directing the control system to reduce the programmed feed rate.

Feed rate override is a percentage function to reduce the programmed feed rate. If the programmed feed rate was 30 inches per minute and the operator wanted 15 inches per minute, the feed rate override dial would be set at 50 percent.

FILE. Organized collection of related data. For example, the entire set of inventory master data records makes up the Inventory Master File.

FIRMWARE. Programs or control instructions that are not changeable (by the user) and that are held in read-only memory (ROM) or another permanent memory device.

FIXED BLOCK FORMAT. A format in which the number and sequence of words and characters appearing in successive blocks is constant.

FIXED CYCLE. See *canned cycle.*

FIXED SEQUENTIAL FORMAT. A means of identifying a word by its location in a block of information. Words must be presented in a specific order, and all possible words preceding the last desired word must be present in the block.

FLOATING ZERO. A characteristic of a machine control unit that permits the zero reference point on an axis to be established readily at any point in the travel.

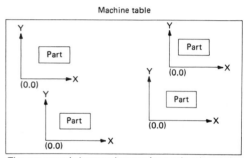

The part or workpiece may be moved to *any* location on the machine table and zero may be established at that point.

FMS. Flexible manufacturing system.

FORMAT (TAPE). The general order in which information appears on the input media, such as the location of holes on a punched tape or the magnetized areas on a magnetic tape.

FULL-RANGE FLOATING ZERO. A characteristic of a numerical machine tool control that permits the zero point on an axis to be shifted readily over a specified range. The control retains information on the location of permanent zero.

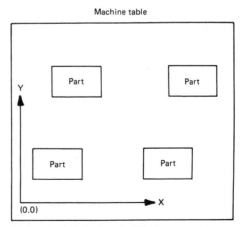

The part or workpiece may be shifted to any position on the machine table, but the actual position of permanent zero remains constant.

GAGE HEIGHT. A predetermined partial re-traction point along the Z-axis to which the cutter retreats from time to time to allow safe XY table travel.

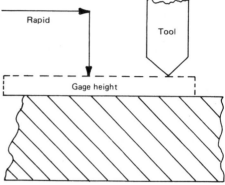

Gage height, usually .100 to .125, is a set distance established in the control or set by the operator. Gage height allows the tool, while advancing in rapid traverse, to stop at the established distance (gage height) and begin feed motion. Without gage height, the tool would rapid into the part causing tool damage or breakage and potential operator injury.

G CODE. A word addressed by the letter G and followed by a numerical code, defining preparatory functions or cycle types in a numerical control system.

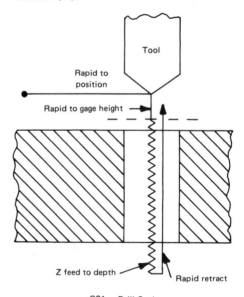

G81 — Drill Cycle

GENERAL PROCESSOR. 1. A computer program for converting geometric input data into the cutter path data required by an N/C machine. 2. A fixed software program designed for a specific logical manipulation of data.

HARD COPY. A readable form of data output on paper.

HARDWARE. The component parts used to build a computer or control system (for example, integrated circuits, diodes, transistors).

HARD-WIRED. Having logic circuits interconnected on a backplane to give a fixed pattern of events.

HIGH-SPEED READER. A reading device that can be connected to a computer or control so as to operate on-line without seriously holding up the computer or control.

INCREMENTAL DIMENSIONING. The method of expressing a dimension with respect to the location of the preceding point in a sequence of points.

INCREMENTAL SYSTEM. A control system in which each coordinate or positional dimension, both input and feedback, is given with respect to the previous position rather than from a common datum point, as in the absolute system.

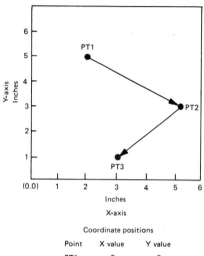

Coordinate positions

Point	X value	Y value
PT1	2	5
PT2	3	-2
PT3	-2	-2

In an incremental system, all points are expressed relative to the preceding point.

INDEX TABLE CODE. A multiple-character code containing the letter B followed by digits. This code determines the position of the rotary index table in degree. (See B-axis)

INHIBIT. To prevent an action or acceptance of data by applying an appropriate signal to the appropriate input.

INPUT. Transfer of external information into the control system.

INPUT MEDIA. 1. The form of input, such as punched cards and tape or magnetic tape. 2. The device used to input information.

INTERCHANGEABLE VARIABLE BLOCK FORMAT. A programming arrangement consisting of a combination of the word address and tab sequential formats to provide greater compatibility in programming. Words are interchangeable within the block. The length of a block varies because words may be omitted. (See *block*)

INTERCHANGE STATION. The position where a tool of an automatic tool-changing machine awaits automatic transfer to either the spindle or the appropriate coded drum station.

INTERFACE. Connection or linkage between software modules that are usually in the same mode.

INTERMEDIATE TRANSFER ARM. The mechanical device in automatic tool-changing machine that grips and removes a programmed tool from the coded drum station and places it into the interchange station, where it awaits transfer to the machine spindle. This device then automatically grips and removes the used tool from the interchange station and returns it to the appropriate coded drum station.

INTERPOLATION. 1. The insertion of intermediate information based on an assumed order or computation. 2. A function of a control whereby data points are generated between given coordinate positions.

INTERPOLATOR. A numerical control system device that performs interpolation.

ISO. International Organization for Standardization.

JOG. A control function that momentarily operates a drive to the machine.

LEADING ZEROES. Redundant zeroes to the left of a number.

Leading zeroes

$$X + \overbrace{00}62500$$

LEADING ZERO SUPPRESSION. See *zero suppression*.

LETTER ADDRESS. The means by which information is directed to different parts of the system. All information must be preceded by its proper letter address (for example, X, Y, Z, M).

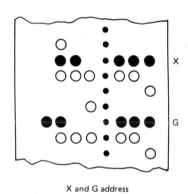

X and G address

An identifying letter inserted in front of each word.

LINEAR INTERPOLATION. A function of a control whereby data points are generated between given coordinate positions to allow simultaneous movement of two or more axes of motion in a linear (straight) path.

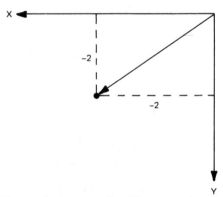

The control system moves X- and Y-axes proportionately to arrive at the destination point.

LOOP TAPE. A short piece of tape, with joined ends, that contains a complete program or operation.

MACHINE CODE. Code obeyed by a computer or microprocessor system with no further translation (normallly written in binary code).

MACHINE LANGUAGE. Basic language of a computer. Programs written in a cryptic machine language that only the computer understands.

MACHINING CENTER. Machine tools, usually numerically controlled, capable of automatically drilling, reaming, tapping, milling, and boring multiple faces of a part. They are often equipped with a system for automatically changing cutting tools.

MACRO. A group of instructions that can be stored and recalled as a group to solve a recurring problem.

An APT macro could be as follows:

```
DRILL1 = MACRO/X, Y, Z, Z1, FR, RR
    GOTO/POINT, X, Y, Z, RR
    GODLTA/–Z1, FR
    GODLTA/+Z1, RR
TERMAC
```

X, Y, Z, Z1, FR, and RR would be variables which would have values assigned when the macro is called into action. The variables would be as follows:

X = X position
Y = Y position
Z = Z position (above work surface)
Z1 = Z feed distance
FR = feed rate
RR = rapid rate

The call statement could be:
CALL/DRILL1, X = 2, Y = 4, Z = .100, Z1 = 1.25, FR = 2, RR = 200

MAGIC-THREE CODING. A feed rate code that uses three digits of data in the F word. The first digit defines the power-of-10 multiplier. It determines the positioning of the floating decimal point. The last two digits are the most significant digits of the desired feed rate.

To program a feed rate of 12 inches per minute in magic-three coding:
1) count the number of decimal places to the left of the decimal. 12 = 2
2) Add magic "3" to the number of counted decimal places. (3 + 2 = 5)
3) write the F word address, the added digit, and the first two digits of the actual feed rate to be programmed. (F512)
4) F512 would be the magic "3" coded feed rate.

This method of feed rate coding is now almost obsolete.

MAGNETIC TAPE. A tape made of plastic and coated with magnetic material. It stores information by selective polarization of portions of the surface.

MANUAL DATA INPUT. A mode or control that enables an operator to insert data into the control system. This data is identical to information that could be inserted by tape.

MANUAL PART PROGRAMMING. The preparation of a manuscript in machine control language and format to define a sequence of commands for use on an N/C machine.

Manual, or hand, programming is programming the actual codes, X and Y positions, functions, etc. as they are punched in the N/C tape.

H001 G81 X+37500 Y+52500 W01

MANUSCRIPT. A written or printed copy, in symbolic form, containing the same data as that punched on cards or tape or retained in a memory unit.

MEMORY. An organized collection of storage elements (for example, disks, drums, ferrite cores) into which a unit of information consisting of a binary digit can be stored and from which it can later be retrieved.

A computer with a 64,000-word capacity is said to have a memory of 64 K.

MIRROR IMAGE. See *axis inversion.*

MODAL. Pertaining to information that is retained by the system until new information is obtained and replaces it.

MODULE. An interchangeable plug-in item containing components.

N/C. (numerical control) The technique of controlling a machine or process by using command instructions in coded numerical form.

NETWORK ARCHITECTURE. Set of rules, standards, or recommendations through which various computer hardware, operating systems, and applications software function together.

NODE. Point in a network where service is provided or used or where communications channels are interconnected.

NULL. 1. Pertaining to no deflection from a center or end position. 2. Pertaining to a balanced or zero output from a device.

NUMERICAL CONTROL SYSTEM. A system in which programmed numerical values are directly inserted, stored on some form of input medium, and automatically read and decoded to cause a corresponding movement in a machine or process.

OFFSET. A displacement in the axial direction of the tool equal to the difference between the actual tool length and the programmed tool length. (Compare with *cutter offset.*)

ON-LINE. Pertaining to peripheral devices operating under the direct control of the central processing unit.

OPEN-LOOP SYSTEM. A control system that has no means of comparing the output with the input for control purposes (that is, no feedback).

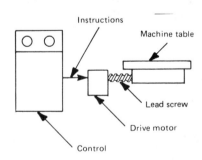

OPERATING SYSTEM. Software that controls the execution of computer programs and that may provide scheduling, input-output control, compilation, data management, debugging, storage assignment, accounting, and other similar functions.

OPTIMIZE. 1. To rearrange the instructions or data in storage so that a minimum number of transfers are required in the running of a program. 2. To obtain maximum accuracy and minimum part production time by manipulation of the program.

OPTIONAL STOP. A miscellaneous function command (M01) similar to Program Stop except that the control ignores the command unless the operator has previously pushed a button to validate the command.

OUTPUT. Data transferred from a computer's internal storage unit to storage or an output device.

OVERSHOOT. The amount by which the motion exceeds the target value. The amount of overshoot depends on the feed rate, the acceleration of the slide unit, or the angular change in direction.

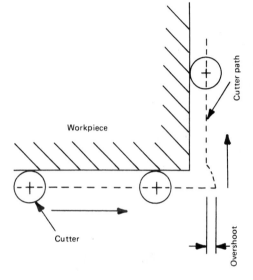

PARABOLA. A plane curve generated by a point moving so that its distance from a fixed second point is equal to its distance from a fixed line.

PARABOLIC INTERPOLATION. Control of cutter path by interpolation between three fixed points, with the assumption that the intermediate points are on a parabola.

PARITY CHECK. 1. A hole punched in one of the tape channels whenever the total number of holes is even, to obtain an odd number (or vice versa, depending on whether the check is even or odd). 2. A check that tests whether the number of ones (or zeroes) in any array of binary digits is odd or even.

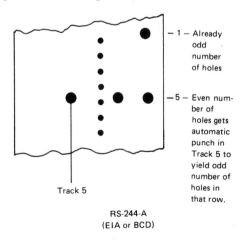

PART PROGRAM. A specific and complete set of data and instructions written in source languages for computer processing or in machine language for manual programming to manufacture a part on an N/C machine.

PART PROGRAMMER. A person who prepares the planned sequence of events for the operation of a numerically controlled machine tool.

PERFORATED TAPE. A tape on which a pattern of holes or cuts is used to represent data.

PLOTTER. A device that draws a plot or trace from coded N/C data input.

POINT-TO-POINT CONTROL SYSTEM. A numerical control system in which controlled motion is required only to reach a given end point, with no path control during the transition from one end point to the next.

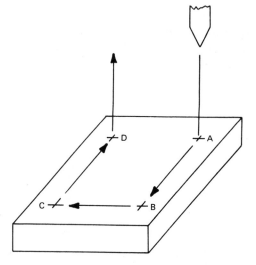

PORT. Point at which data can enter or leave a network, such as the serial or parallel ports in the backs of PCs.

POSITIONING/CONTOURING SYSTEM. A type of numerical control system that has the capability of contouring, without buffer storage, in two axes and positioning in a third axis for such operations as drilling, tapping, and boring.

POSITIONING SYSTEM. See *point-to-point control system.*

POSITION READOUT. A display of absolute slide position as derived from a position feedback device (transducer) normally attached to the lead screw of the machine. See *command readout.*

POSTPROCESSOR. The part of the software that converts the cutter path coordinate data into a form that the machine control can interpret correctly. The cutter path coordinate data is obtained from the general processor based on all other programming instructions and specifications for the particular machine and control.

PREPARATORY FUNCTION. An N/C command on the input tape that changes the mode of operation of the control (generally noted at the beginning of a block by the letter G plus two digits.) (See *G code*)

Some preparatory functions are:

G84	—	tap cycle
G01	—	linear interpolation
G82	—	dwell cycle
G02	—	circular interpolation — clockwise
G03	—	circular interpolation — counter clockwise

PROCESSOR. A program that translates a source program into object language.

PROGRAM. A sequence of steps to be executed by a control or a computer to perform a given function.

PROGRAMMED DWELL. A delay in program execution for a programmable length of time.

PROGRAMMER. See *part programmer*.

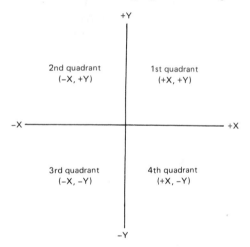

Manual part programming instructions:

```
H001  G81  X+123750   Y+62500    W01
N002        X+105000
N003                   Y+51250    M06
```

Computer part programming instructions:

```
TLRGT,  GORGT/HL3, TANTO, C1
         GOFWD/C1, TANTO, HL2
         GOFWD/HL2, PAST, VL2
```

PROGRAM STOP. A miscellaneous function command (M00) to stop the spindle, coolant, and feed after completion of the dimensional move commanded in the block. To continue with the remainder of the program, the operator must push a button.

PROTOCOL. Set of messages with specific formats and rules for exchanging the messages.

QUADRANT. Any of the four parts into which a plane is divided by rectangular coordinate axes in that plane.

```
                    +Y
                     |
   2nd quadrant      |   1st quadrant
   (–X, +Y)          |   (+X, +Y)
                     |
-X ------------------+------------------ +X
                     |
   3rd quadrant      |   4th quadrant
   (–X, –Y)          |   (+X, –Y)
                     |
                    –Y
```

RANDOM. Not necessarily arranged in the order of use, but having the ability to select from any location in, and in any order from, the storage system.

RANDOM-ACCESS MEMORY (RAM). Memory device that can be written and read under program control and that is often used as a scratchpad memory by the control logic or by the user for his or her data.

RAPID. To position the cutter and workpiece into close proximity with one another at a high rate of travel speed, usually 400 to 1200 inches per minute (IPM) or more, before the cut is started.

READER. A pneumatic, photoelectric, or mechanical device used to sense bits of information on punched cards, punched tape, or magnetic tape.

READ-ONLY MEMORY (ROM). Memory device containing information that is fixed and can only be read and not changed.

REGISTER. An internal array of hardware binary circuits for temporary storage of information.

REPEATABILITY. Closeness of, or agreement in, repeated measurements of the same characteristics for the same method and the same conditions.

RESET. To return a register or storage location to zero or to a specified initial condition.

ROUTINE. Set of functionally related instructions that directs the computer to carry out a desired operation. A subdivision of a program.

ROW (TAPE). A path perpendicular to the edge of the tape along which information may be stored by the presence or absence of holes or magnetized areas. A character would be represented by a combination of holes.

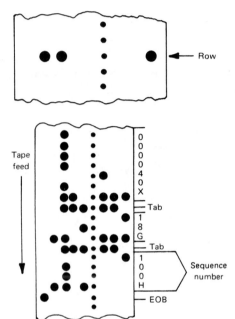

SEQUENCE NUMBER (CODE WORD). A series of numerals programmed on a tape or card and sometimes displayed as a readout; normally used as a data location reference or for card sequencing.

SEQUENCE READOUT. A display of the number of the block of tape being read by the tape reader.

SEQUENTIAL. Arranged in some predetermined logical order.

SERVER. Module or set of modules that performs a well-defined service, such as remote file access or gateway communication, on behalf of another module.

SIGNIFICANT DIGIT. A digit that must be kept to preserve a specific accuracy or precision.

Significant digits

$$X + \underbrace{00\overbrace{525}00}$$

X + 0052500

Insignificant digits

SIMULATE. To use one system to represent the functioning of another; that is, to represent a physical system by the execution of a computer program or to represent a biological system by a mathematical model.

SLOW-DOWN SPAN. A span of information having the necessary length to allow the machine to decelerate from the initial feed rate to the maximum allowable cornering feed rate that maintains the specified tolerance.

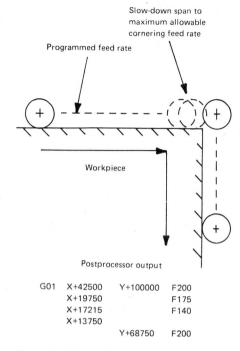

Postprocessor output

G01	X+42500	Y+100000	F200
	X+19750		F175
	X+17215		F140
	X+13750		
		Y+68750	F200

SOFTWARE. Instructional literature and computer programs used to aid in part programming, operating, and maintaining the machining center.

Examples of software programs are:

APT
FORTRAN
COBOL
RPG

SOURCE PROGRAM. Computer program written in a symbolic programming language (for example, an assembly language program, FORTRAN program, or COBOL program). A translator is used to convert the source program into an object program that can be executed on a computer.

SPAN. A certain distance or section of a program designated by two end points for linear interpolation: a beginning point, a center point, and an end point for circular interpolation; and two end points and a diameter point for parabolic interpolation.

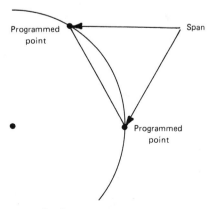

One linear interpolation span

SPINDLE SPEED (CODE WORD). A multiple-character code containing the letter S followed by digits. This code determines the RPM of the cutting spindle of the machine.

STATEMENT. Agreed-upon arrangement of words and/or data accepted by a system to command a particular computer function.

STORAGE. A device into which information can be introduced, held, and then extracted at a later time.

SUBROUTINE. Sequence of callable computer programming statements or instructions that performs frequently required operations (also called a macro).

TAB. A nonprinting spacing action on tape preparation equipment. A tab code is used to separate words or groups of characters in the tab sequential format. The spacing action sets typewritten information on a manuscript into tabular form.

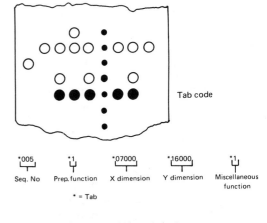

Tab code

TAB SEQUENTIAL FORMAT. Means of identifying a word by the number of tab characters that precede the word in a block. The first character of each word is a tab character. Words must be presented in a specific order, but all characters in a word, except the tab character, may be omitted when the command represented by that word is not desired.

The tab sequential format is, for the most part, obsolete.

TAPE. A magnetic or perforated paper medium for storing information.

TAPE LAGGER. The trailing end portion of a tape.

TAPE LEADER. The front, or lead, portion of a tape.

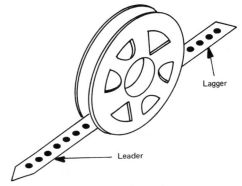

Reel tapes should have a leader and lagger of approximately three feet with just sprocket holes for tape loading and threading purposes.

TERMINAL. Unit in a network, such as a modem or telephone, at which data can be either sent or recieved.

TOOL FUNCTION. A tape command identifying a tool and calling for its selection. The address is normally a T word.

T06 would be a tape command calling for the tool assigned to spindle or pocket 6 to be put in the spindle.

TOOL LENGTH COMPENSATION. A value input manually, by means of selector switches, to eliminate the need for preset tooling; allows the programmer to program all tools as if they are of equal length.

TOOL OFFSET. 1. A correction for tool position parallel to a controlled axis. 2. The ability to reset tool position manually to compensate for tool wear, finish cuts, and tool exchange.

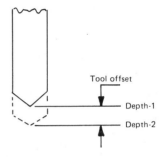

Tool offsets are used as final adjustments to increase or decrease depths due to cutting forces and tool deflection. In this case, a tool offset could be used to increase the drill depth from depth-1 to depth-2.

TOPOLOGY. Physical arrangement and relationship of interconnected nodes and lines in a network.

TRAILING ZERO SUPPRESSION. See *zero suppression*.

TURN KEY SYSTEM. A term applied to an agreement whereby a supplier will install an N/C or computer system such that he or she has total responsibility for building, installing, and testing the system.

USASCII. United States of America Standard Code for Information Interchange. See *ASCII*.

VARIABLE BLOCK FORMAT (TAPE). A format that allows the quantity of words in successive blocks to vary.

Same as word address. Variable block means the length of the blocks can vary depending on what information needs to be conveyed in a given block. See *block*.

VECTOR. A quantity that has magnitude, direction, and sense; it is represented by a directed line segment whose length represents the magnitude and whose orientation in space represents the direction.

VECTOR FEED RATE. The feed rate at which a cutter or tool moves with respect to the work surface. The individual slides may move more slowly or quickly than the programmed rate, but the resultant movement is equal to the programmed rate.

VOLATILE STORAGE. Storage medium in which data connot be retained without continuous power dissipation.

WORD. An ordered set of characters; the normal unit in which information may be stored, transmitted, or operated upon.

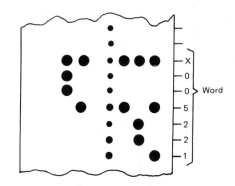

WORD ADDRESS FORMAT. The specific arrangement of addressing each word in a block of information by one or more alphabetical characters that identify the meaning of the word. (See *word*)

WORD LENGTH. The number of bits or characters in a word. (See *word*)

X-AXIS. Axis of motion that is always horizontal and parallel to the work-holding surface.

Y-AXIS. Axis of motion that is perpendicular to both the X- and Z-axes.

Z-AXIS. Axis of motion that is always parallel to the principal spindle of the machine.

ZERO OFFSET. A characteristic of a numerical control machine tool, permitting the zero point on an axis to be shifted readily over a specified range. The control retains information on the location of the permanent zero. (See *full range floating zero* and *floating zero*)

ZERO SHIFT. A characteristic of a numerical machine tool control permitting the zero point on an axis to be shifted readily over a specified range. The control does not retain information on the location of the permanent zero. (See *floating zero*. Consult Chapter 3 for additional details)

ZERO SUPPRESSION. Leading zero suppression: the elimination of insignificant leading zeroes to the left of significant digits, usually before printing. Trailing zero suppression: the elimination of insignificant trailing zeroes to the right of significant digits, usually before printing.

Leading zero suppression

X + 0043500

Insignificant digits

Could be written as:

X + 43500

Trailing zero suppression

X + 0043500

Insignificant digits

Could be written as:

X + 00435

INDEX